SpringerBriefs in Computer Science

Series Editors
Stan Zdonik
Peng Ning
Shashi Shekhar
Jonathan Katz
Xindong Wu
Lakhmi C. Jain
David Padua
Xuemin (Sherman) Shen
Borko Furht
V.S. Subrahmanian
Martial Hebert
Katsushi Ikeuchi
Bruno Siciliano
Sushil Jajodia

For further volumes:
http://www.springer.com/series/10028

Mi Wen • Rongxing Lu • Xiaohui Liang
Jingsheng Lei • Xuemin (Sherman) Shen

Querying over Encrypted Data in Smart Grids

 Springer

Mi Wen
College of Computer Science
 and Technology
Shanghai University of Electric Power
Shanghai, China

Rongxing Lu
School of Electrical and Electronics
 Engineering
Nanyang Technological University
Nanyang, Singapore

Xiaohui Liang
Institute for Security, Technology,
 and Society
Dartmouth College
Hanover, NH, USA

Jingsheng Lei
College of Computer Science
 and Technology
Shanghai University of Electric Power
Shanghai, China

Xuemin (Sherman) Shen
Department of Electronic and Computer
 Engineering
University of Waterloo
Waterloo, ON, Canada

ISSN 2191-5768   ISSN 2191-5776 (electronic)
ISBN 978-3-319-06354-6   ISBN 978-3-319-06355-3 (eBook)
DOI 10.1007/978-3-319-06355-3
Springer Cham Heidelberg New York Dordrecht London

Library of Congress Control Number: 2014937399

Printed on acid-free paper

Springer is part of Springer Science+Business Media (www.springer.com)

Education Commission No. 13ZZ131, No. 14YZ129; Foundation Key Project of Shanghai Science and Technology Committee No. 12JC1404500, and Project of Shanghai Science and Technology Committee No. 12510500700.

Shanghai, China   Mi Wen
Nanyang, NH, Singapore   Rongxing Lu
Hanover, USA   Xiaohui Liang
Shanghai, China   Jingsheng Lei
Waterloo, ON, Canada   Xuemin (Sherman) Shen

Preface

How to query over encrypted data is an important and challenging problem for the smart grid, especially when encryption is required to protect data privacy for decision making. Though querying encrypted data has been well researched in both cryptography and database communities, little attention has been paid to the multidimensional characteristics of metering data in the smart grid. Therefore, the existing query techniques cannot be directly applied to the smart grid. In this brief, we provide a comprehensive study of encrypted data query in the smart grid. Three kinds of queries are introduced, namely, equality query, conjunctive query and range query. Detailed security and performance analysis are also provided. Future research directions are suggested. We hope this brief could be a useful reference for graduate students and professionals who are interested in encrypted data query in smart grid.

In Chap. 1, we give an overview of the concept of the smart grid architecture and discuss the security challenges of the smart grid and the existing encrypted data query techniques. In Chap. 2, we present an efficient Searchable Encryption Scheme for Auction (SESA). Specifically, public key encryption with keyword search is used to enable the energy sellers to inquire potential winner from the auction server while preserving the privacy of the energy buyers. In addition, an extension of SESA is proposed to support detailed filtering of the bids. In Chap. 3, we address the conjunctive query problem in the smart grid. An Efficient Conjunctive Query (ECQ) scheme is proposed to support conjunctive keywords query on multiple dimensions. To resolve the data privacy problem in financial auditing for the smart grid, we introduce a novel privacy-preserving range query scheme over encrypted metering data, named PaRQ, in Chap. 4. Finally, we draw conclusions and outline future research directions in Chap. 5.

We would like to thank our BBCR colleagues for their valuable comments on the brief. We also would like to thank the Springer editors and staff for their great help in getting this brief published. This research work was supported by the National Natural Science Foundation of China under Grant No. 61373152, No. 61272437 and No. 61202369; NSERC, Canada; Innovation Program of Shanghai Municipal

Contents

Chapter 1
Introduction

Smart grid, envisioned as an indispensable power infrastructure, is featured by real-time and two-way communications. By installing smart meters at users' houses, the smart grid can collect real-time data about power consumption by residential users. The amount of data generated by smart meters and intelligent sensors in smart grid will experience explosive growth in the next few years. According to a recent report from SBI Energy, the volume of smart grid data that will have to be managed by utilities is going to surge from 10,780 terabytes (TB) in 2010 to over 75,200 TB in 2015. How to query and mine this massive and heterogeneous power system raw data to support decision making and ensure reliability will be very critical for smart grid.

For security concern, most of the metering data are encrypted by cryptographic algorithms. Consider that requesters, such as utility companies or marketing managers, are tasked with querying power system data by user types and/or dates, etc. Thus, how to query over encrypted data is critical for the smart grid. In the existing proposals, the problem of querying encrypted data has been deeply researched in both cryptography and database communities, but few are researched in the smart grid. In this chapter, we first introduce the smart grid architecture and security topics in the smart grid, then we focus on designing query techniques over encrypted data and the security primitives.

1.1 Smart Grid Architecture

Smart grid is the next generation power grid system which allows decentralized two-way transmission and reliability and efficiency driven response [1]. More importantly, with the integration of advanced computing and communication technologies, the smart grid is expected to greatly enhance efficiency and reliability of future power systems with renewable energy resources, as well as distributed

M. Wen et al., *Querying over Encrypted Data in Smart Grids*, SpringerBriefs in Computer Science, DOI 10.1007/978-3-319-06355-3_1, © The Author(s) 2014

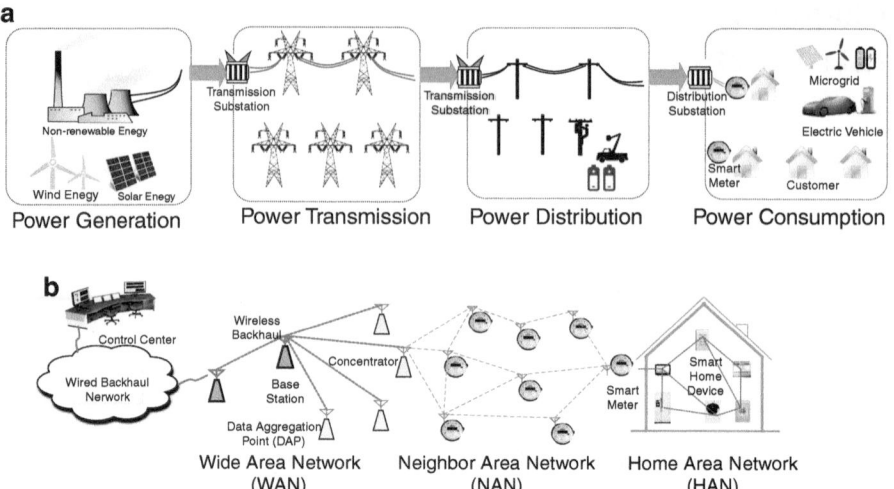

Fig. 1.1 Smart grid architecture. (**a**) Power system layer. (**b**) Communications layer

intelligence and demand response. According to NIST's conceptual model [2], the smart grid consists of seven logical domains: Bulk Generation, Transmission, Distribution, Customer, Markets, Service Provider and Operations. The first four feature the two way power and information flows. The last three feature information collection and power management in the smart grid. More detailed discussion on smart grid domains can be found in [3].

In order to interconnect all these domains, the communication network must be highly-distributed and hierarchical. As shown in Fig. 1.1, we represent the smart grid communication network as a hybrid and hierarchical network, including the backbone network and millions of local-area networks. By building up Smart Grid Communications Network (SGCN), the smart grid systems can support all identified smart grid functionalities such as Advanced Metering Infrastructure (AMI), Demand Response (DR) etc. One of the important components of SGCN is smart meter which is a two-way communication device deployed at consumers premise and can record power consumption periodically. With smart meters, the real-time information about grid operations and status at a control center can be reported to the smart grid systems, through a reliable communication network deployed in parallel to the power transmission and distribution grid, as shown in Fig. 1.1. Thus, we depict the smart grid architecture into two layers: Power system layer and Communication layer.

1.1.1 Power System Layer

The operation of power system layer involves four aspects: power generation using renewable resources (e.g., solar, wind, geothermal, fossil-fuel); power transmission, as shown in Fig. 1.1(a), by balancing the demand of power in different regions and different customers; power distribution by monitoring the power consumption by customers using smart meters [4]. For clarity, the power, which is generated at the power plants, is supplied to the consumers via two components. The first component is the transmission substation at/near the power plant. The second component comprises a number of distribution substations.

The transmission substation delivers power from the power plant over high voltage transmission lines (usually over 230 kV) to the distribution substations, which are located at different regions. The distribution substations transform the electric power into medium voltage level and then distribute it to the building feeders. The building feeders convert the medium voltage level into a lower one for consumers' use [5]. Smart meters are devices acting as an interface between the power distribution and the home area network. They record consumption of electric energy in intervals of an hour or less and communicates that information at least daily back to the utility for monitoring and billing purposes. Smart meters enable two-way communication between the meter and the smart grid control systems. Unlike home energy monitors, smart meters can gather data for remote reporting.

1.1.2 Communications Layer

To explore the smart grid topology from communication point of view, the communication framework is divided into a number of hierarchical networks. The transmission substation located at/near the power plant, and the control centers (CC) of the distribution substations are connected with one another in a meshed network. This mesh network is considered to be implemented over optical fiber technology. The SGCN for the lower distribution network is divided into a number of hierarchical networks comprising Wide Area Network (WAN), Neighborhood Area Network (NAN) and Home Area Network (HAN), as shown in Fig. 1.1(b).

Two-way communication and Smart metering are the two important features of SGCN. Each individual domain is comprised of important smart grid elements that are connected to each other through two-way communications. For the customer domain, the smart meters control and manage the flow of electricity to and from the customers and provide energy information about energy usage and patterns. Each customer has a discrete domain comprised of electricity premise and two-way communications networks. Smart metering essentially involves an electronic power meter supplemented by full remote control, diagnostics, power peak and consumption analysis, anti-tampering mechanisms, fault alert, time-variable tariffs, and many more possibilities. Using power-line communication or other wired and wireless technologies to connect the meter to the service provider enables all of the above features to be feasible and compatible with future smart-grid protocols.

For simplicity, each distribution substations is considered to cover only one neighborhood area network (NAN). If there are n distribution substations covering n neighborhoods, these n NANs are composed of a WAN. On the other hand, every NAN contains a number of HANs. In addition, there are advanced meters called smart meters deployed in the smart grid architecture which comprise AMI for enabling an automated, two-way communication between the utility meter and the utility provider. The smart meters are equipped with two interfaces, namely for reading power and for communication gateway. The smart meters and distribution substations are used in referred to as HAN gateway, NAN gateway and WAN gateway, respectively. Therefore, the WAN, NAN, and HAN are formed in a hierarchical manner.

1.2   Security Challenges in SGCN

With the smart meters deployed in the smart grid system, the information and communication technology (ICT) will be developed to support novel communication and control functions. Unfortunately, this additional dependency also expands the risk from cyber attacks to information or data [6]. It is not only from terrorist attacks, but also from customers and building authorities who can tamper with various devices. As security challenges mainly come from malicious cyber attacks via communication networks, it is essential to understand potential vulnerabilities in the smart grid under network attacks.

For example, pricing information and control actions are transmitted via the communication network. Various attacks such as eavesdropping, information tampering, and malicious control command injection in the communication network, would impose serious threat on secure and stable smart grids operation [7]. Moreover, smart grid is an attractive target for various hackers with diversified motivations, e.g. unethical customers may want to modify their meter readings to evade the electric charge; malicious users are able to extract the behaviors of household by eavesdropping the communications of smart meters, called non-intrusive appliance load monitoring (NILM); vicious terrorists want to inject the false data or command to disrupt the grid [8, 9]. The U.S. National Institute of Standards and Technology noticed the importance of the security of the smart grid, and it laid out the guidelines for developers and policy makers, covering cyber security requirements of the smart grids that should be included from the beginning of the development process [10].

In this section, we provide an overview of attacks towards the smart grid. We first introduce the security objectives in the smart grid communication network, then classify the attacks into four categories. Finally, we analyze the existing countermeasures to these attacks.

1.2.1   Security Objectives in SGCN

The Smart Grid communication network is a mission critical network for information exchange in communication networks [3]. To ensure secure and reliable operation, it is essential to understand what are the security objectives before providing comprehensive countermeasures in the context of energy delivery and management. Here, we describe four high-level Smart Grid security objectives by referencing [11].

- Availability: Ensuring timely and reliable access to and use of information is of the most importance in the Smart Grid. This is because a loss of availability is the disruption of access to or use of information, which may further undermine the power delivery.
- Integrity: Guarding against improper information modification or destruction is to ensure information non repudiation and authenticity. A loss of integrity is the unauthorized modification or destruction of information and can further induce incorrect decision regarding power management.
- Confidentiality: Preserving authorized restrictions on information access and disclosure is mainly to protect personal privacy and proprietary information. This is in particular necessary to prevent unauthorized disclosure of information that is not open to the public and individuals.
- Privacy: Protecting the expanded information, particularly from energy consumers and other individuals, from being identified based on the signatures they exhibit in electric information at the meter. Especial when collections occur with great frequency as opposed to traditional monthly meter readings, this more detailed information expands the possibility of intruding on consumers' and other individuals' personal privacy expectations.

From the perspective of system reliability, availability, integrity and privacy are the most important security objectives in the Smart Grid. The first three ones are deeply studied by researchers, while the privacy is rarely being discussed. When the information is transmitted through different domains; how to protect the privacy is very challenging and critical in smart grid communication networks, particularly in systems involving interactions with customers, such as information query in demand response and AMI networks.

1.2.2   Attacks in SGCN

In general, security attacks can be classified into four categories: eavesdroppers and traffic analyzer, data integrity attack, privacy attack and denial of service (DoS) attack.

- Eavesdroppers and Traffic Analyzers: By eavesdropping on wireless communication channels, an attacker at a home-area network could possibly gain private information even if the information was encrypted [11]. Therefore, strong data encryption and secret key management schemes must be enforced for any communication in the AMI network to prevent attacks from deducing the secret key out of a large amount of network data samples.

- Data Integrity Attack: Data integrity cyber attacks that consist of a set of compromised power meters whose readings are altered by the attacker [12]. Cyber attacks whose compromised meter readings are consistent with the physical power flow constraints are called unobservable. Unobservable attacks require coordination compromised meter readings must be carefully orchestrated to fall on a low dimensional manifold in order for the attack to be unobservable. Unobservable attacks will pass any bad data detection algorithm. Such attacks can cause significant errors in state estimation algorithms, which can mislead system operators into making potentially catastrophic decisions [13]. Liu et al. [14] have recently shown that many power systems commonly admit unobservable attacks involving a relatively small number of power meters, and consequently the degree of coordination necessary is modest. This surprising result has led to a flurry of activity in the power system cyber security research community [15].

  In addition, similar to the Internet and data collection in sensor networks, conventional relay or man-in-the-middle attacks can be possibly launched in the AMI network to inject falsified data during the communication process [16]. To this end, end-to-end encryption and authentication schemes are required to eliminate such attacks in the AMI network.

- Privacy Attack: Privacy attack aims to learn/infer users' private information by analyzing electricity usage data. In smart grid, electricity usage information is collected multiple times per hour by smart meters so as to obtain fine-grained information about the grid status and improve grid operation efficiency [17]. The detailed information may easily reveal customers' physical activities. For example, in a residential setting, lack of electricity use for stove and microwave during a certain time period indicates that the home is not occupied [1]. Using this information, physical attacks like robbery can be planned when nobody is at home. Clearly, such privacy-sensitive information must be protected from unauthorized access not only during the data depositing domain, but also during the data query domain.

- DoS Attack: The objectives of DoS attack are to use up or overwhelm the communication and computational resources of the smart grid, resulting in delay or failure of data communications. For example, an adversary may flood a control center with false information at very high frequency such that the control center spends most of the time verifying the authenticity of the information and is not able to timely respond to legitimate network traffic [17]. In fact, a recent work in [18] has already pointed out that attackers can focus on jamming real-time price signals transmitted in wireless home-area networks to effectively result in denial-of-service and dysfunction of the entire demand respond system. Communication and control in smart grid are time critical. A delay of a few seconds may cause irreparable damage to the national economy and homeland security. A network availability attack must be handled effectively.

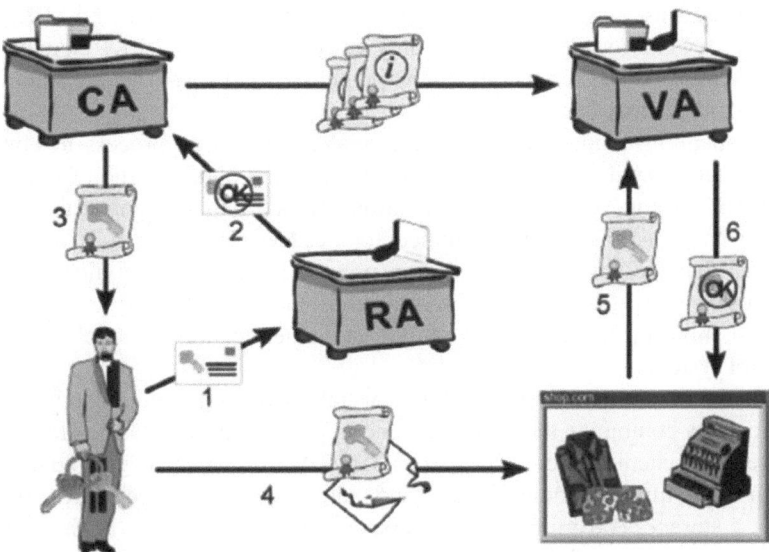

Fig. 1.2 PKI procedure [19]

1.2.3 Coutermeasures in SGCN

There exist several security techniques for defending the above attacks in smart grid communication networks. These techniques are to be used in combination to provide full protection against existing or future sophisticated malicious attacks.

- Public Key Infrastructure (PKI): The PKI infrastructure has been used to protect smart grids [19]. This infrastructure should provide certification to various components and devices in the network. Specific certification policies need to be issued. The basic steps in utilizing a PKI, as shown in Fig. 1.2, are as follows. First, in order to communicate with a secure resource (SR), the device (i.e., a certificate subject) begins to send a certificate signing request to a registration authority (RA). The RA performs a validation function check to determine whether the requested bindings are correct or not. If the requested bindings are correct, RA signs the certificate signing request and forwards it to the certificate authority (CA), which then issues a certificate. CA will issue the certificate and let the validation authority (VA) know the certificate information of the certificated subject at the same time. Later, when a device wants to access a secure resource, it transmits a certificate to the SR. The SR validates the certificate, typically by requesting the certificate status from a VA. Eventually, the VA will reply a positive response if the certificate is valid [20].

  While PKI is known for being complex, the authors in [20] suggested that many of the items responsible for the complexity can be significantly reduced by including the following four major technical elements, e.g., PKI standards, automated trust anchor security, certificate attributes, and smart grid PKI tools. Ma et al. [21] demonstrated that only by including these PKI elements into the overall security architecture, a comprehensive and cost-effective solution for smart grid security can be achieved.

- Authentication: Authentication is a crucial identification process to eliminate attacks targeting data integrity. Intuitively, design of authentication for the smart grid can leverage existing authentication protocols in conventional networks, which have been extensively studied for decades. However, it is pointed out in [22] that the authentication design process is prone to significant errors if adequate care is not taken for power systems. Few of recent works are designed toward this objective, i.e., fast multicast authentication protocols for power control systems.

  Fouda et al. [5] proposed a lightweight two-step mutual authentication protocol by combining the public key encryption scheme and Diffie-Hellman key agreement scheme. Li et al. [23] considered that in the smart grid field devices often have limited storage space, and addressed the authentication problem from a storage load minimization perspective. Authentication is also achieved using new variant of the one-time signature (OTS) [23] and signed message authentication code (SIGMA) [24]. Though these authentication protocols for the smart grid can protect data integrity, rare of them can meet the requirements of the smart grid at the same time, such as high efficiency, tolerance to faults and attacks and support of multicast.

- Access Control: Since the data at the substations is sensitive and needs to be well protected, most of them should be encrypted and stored securely in the data repository. However, different tasks are performed by separate users; for example, the electrical and maintenance board will monitor the network, the costs calculation and analysis is done by the auditing unit, and network planners/researchers can be involved to predict future behavior [25]. All information must be sent only to the users responsible for a specific job. Hence, there is a need for fine-grained access control, letting information only can be accessed by authorized users.

  Attribute Based Encryption (ABE) is another kind of access control mechanism. ABE is first proposed by Sahai and Waters in 2006 [26]. The main idea is to distribute attributes to receivers and attributes to senders so that only the receivers with matching attributes structure can access the data. Data is encrypted using attribute based keys, which are distributed by a central key distribution center (KDC). This one is well-known as Key Policy ABE (KP-ABE). Another variant of ABE is Ciphertext-Policy ABE (CP-ABE) [27], where the private key of the user is associated with a set of attributes, and the encrypted message specifies an access policy over the attributes. To decrypt a given ciphertext, the user needs his attributes to satisfy the access policy, which is specified within the ciphertext. An example of how CP-ABE can be applied to the smart grid communication can be found in [28], where CP-ABE gives selective access to user data stored in the smart grid data repositories. However, the efficiency of ABE should be improved in the smart grid.

- Privacy Preservation: Since in a regional network, smart metering data flows to the control center through community gateways, we reasonably suggested that community gateways aggregate received raw smart metering data and forward them in an integrated compact form to the control center. Each smart meter encrypts its readings using a secret key shared with the control center. Community gateways can not be able to access the content of metering data. To enable community gateways to perform data aggregation in this case, [29] proposes a secure aggregation protocol for the wireless-based AMI network. In [29], end-to-end security is achieved via a shared secret between the source and the destination; hop-by-hop security is enforced at the physical/MAC layers via pairwise keys between a node and its next-hop neighbors. There are still several issues associated with current approaches. For example, both [30] and [29] assume that all nodes are trustworthy and there is no attack along the aggregation path.

Another recent work [1] suggests that smart meters apply a homomorphic encryption technique. In this technique, a specific linear algebraic manipulation toward the plaintext is equivalent to another one conducted on the ciphertext. This unique feature allows community gateways to perform summation and multiplication based aggregation on received metering data without decrypting them. It can partly resolve the drawbacks of the above, but it is not efficient if there is large scale of data. Accordingly, secure aggregation protocols for the smart grid need to protect data integrity and confidentiality, and also be resilient to malicious attacks and achieve efficiency.

1.3 Existing Techniques for Encrypted Data Query

Currently, there is not much work which address the encrypted data query problem in a smart grid. Existing literature focus either on authentication [5] or privacy protection [31] or aggregation [1]. They all focus on how to protect or encrypt the data in a secure manner, but none of them take how to query over the encrypted data into account for the smart grid. In fact, there are requirements to do so. The cyber security working group in the NIST Smart Grid interoperability panel has recently released a guidelines for the smart grid cyber security: Privacy and the Smart Grid [11]. It pointed that the metering data with high frequency could release the customers' privacy. Cryptographic countermeasures, such as encryption algorithms should be used to protect the data privacy. However, when these encrypted data is used for decision making and price predicting etc. how to query these encrypted data by user types and/or dates is critical for the smart grid.

Thus, there is a need to design encrypted data query techniques to find useful information from those ciphtexts. Meanwhile, the data confidentiality, integrity, privacy should also be achieved. In this manner, smart meters can send their data in a secure way and utility companies or marketing managers can get what they need from the encrypted data. The problem of querying encrypted data has been deeply researched in both cryptography and database communities [32]. The specific scheme that may be applicable depends on the type of data and the type of query that needs to be supported. We describe some of the techniques proposed as follows.

1.3.1 Order-Preserving Encryption

An ideal cryptographic scheme for supporting queries would be one that allows comparison-predicate evaluation over encrypted data directly. Agrawal et al. [33] were the first ones to propose an order-preserving encryption scheme for numeric data. Such a scheme allows the server to execute range queries directly on the encrypted representation as well as maintain indexes which would enable efficient access of the encrypted data. Recently, Boldyreva et al. [34] gave a formal security analysis and a relatively efficient implementation of such an order-preserving encryption scheme. They found the disadvantage of order-preserving schemes is that the encryption needs to be deterministic (i.e., all encryption of a given plaintext should be identical). As a result, it reveals the frequency of each distinct value in the dataset and hence susceptible to statistical attacks. Both these techniques suffer from the problem that the adversary is able to progressively improve his estimate of the true value of a ciphertext by evaluating more predicates against the ciphertext (i.e., as number of queries increase).

1.3.2 Searchable Encryption Techniques

Recently, Boneh et al. [35] introduced the concept of public key encryption with keyword search (PEKS), which can support equality query over encrypted data. Then Park et al. [36] proposed the notion of public key encryption with conjunctive field keyword search to support conjunctive query. Shi et al. [37] proposed a searchable encryption scheme that supports multidimensional range queries over encrypted data (MRQED). The technique utilizes an interval tree structure to form a hierarchical representation of intervals along each dimension and stores multiple ciphertexts corresponding to a single data value on the server (one corresponding to each level of the interval tree). Consider the case when the scheme is applied to a single dimensional dataset with values belonging to a domain of size N. The scheme not only generates a ciphertext representation that is $O(logN)$ times the actual dataset (due to multiple encrypted representation), each search query has a complexity of $O(KlogN)$ where K is the size of the dataset. The performance only gets worse as the dimension increases each query requires an $O(dKlogN)$ time to execute over a d-dimensional space. The scheme is proven secure under the selective identity security model [35] similar to that used for identity based encryption techniques.

A binary string-based encoding of ranges was introduced in [38]. Similar to [37], they show that they can represent each range as a set of prefixes (this is same as saying that a set of nodes in the interval tree can represent a range as done in [37]). Then, using the bit-flipping- based encryption scheme [39] they modify the encoding function (i.e., flip the bits corresponding to a subset of branches in the interval tree). This modified encoding function is called the ciphertext tree which is then used for encoding the numeric values and also translating a range query into appropriately represented set of prefixes. There are severe performance limitations for the techniques described above. For instance, public key encryption based schemes in [40] are substantially slower than symmetric key encryption algorithms. Further, neither one of these techniques admit any form of indexing for faster access. This leads to inefficient access mechanism as the server has to check the query predicate against each record, i.e., make a complete scan of the whole table. This could be prohibitively expensive for large tables. Additionally, like the order preserving encryption scheme, the techniques of [37, 40] are also susceptible to the value-localization problem described above.

1.3.3   Special Data Structure Traversal

Recently, Vimercati et al. [41] created a bucket based on the B+tree index. In order to hide the access patterns, they run fake queries. In addition, they buffer the first few levels of the tree on the client side. Finally, they shuffle from time to time. However, this technique is trying to solve a different problem outsourcing a B+tree securely and obliviously traversing this data structure to retrieve data. While our technique tries to prevent approximate leakage under certain attack scenarios (e.g., the adversary may learn the distribution of values within a bucket), this paper does not analyze the security under any specific attacks. Also, how the technique will scale with dimensions or be applied to multidimensional data structures remains unclear at this point. Furthermore, practical oblivious data structure traversals techniques could potentially be integrated into a secure data outsourcing framework.

1.3.4   Data Partitioning Problems

Many optimization problems in computer science and related areas can be modeled as a multidimensional partitioning problem, e.g., data anonymization for privacy-preserving data publication [42] etc. In a typical setting, the partitioning problem consists of a set of data elements along with an objective function and a set of constraints. The data element is represented as points in some suitably chosen d-dimensional space and the optimization goal is to partition this set into a number of bins such that the value of the objective function is minimized (or maximized) and at the same time, all the constraints are met. The objective function is often an aggregate function over the resulting bins. For instance, in the problem of multidimensional histogram creation, if data elements are associated with a weight equal to their frequency in the dataset, the objective is to minimize the total sum of square of deviations from the mean value in each bin. Such a partition tends to make within-bin distribution more uniform and leads to high-quality histograms for density estimation problems. The constraints define the subset of partitioning schemes that are admissible (feasible) for the given application.

Hore B. et al. [32] developed a bucketization procedure for answering multidimensional range queries on multidimensional data. For a given bucketization scheme, we derive cost and disclosure-risk metrics that estimate client's computational overhead and disclosure risk respectively. Given a multidimensional dataset, its bucketization is posed as an optimization problem where the goal is to minimize the risk of disclosure while keeping query cost (client's computational overhead) below a certain user-specified threshold value.

1.4 Security Primitives

In this section, we will briefly describe the basic definition and properties of bilinear pairings, Public-key Encryption with Keyword Search (PEKS) and Hidden Vector Encryption (HVE).

1.4.1 Bilinear Pairing

Bilinear pairing is an important cryptographic primitive. Let G_1 and G_2 be two cyclic multiplication groups of prime order q. Let a and b be elements of Z_q^*. We assume that the discrete logarithm problem (DLP) in both G_1 and G_2 are hard. g is a generator of G_1. A bilinear pairing is a map $e : G_1 \times G_1 \rightarrow G_2$ with the following properties:

1. Bilinear: $e(g^a, h^b) = e(g, h)^{ab}$ for any $(g, h) \in G_1^2$;
2. Non-degenerate: $e(g, h) \neq 1_{G_2}$ whenever $g, h \neq 1_{G_1}$;
3. Computable: There is an efficient algorithm to compute $e(g, h) \in G_2$ for all $(g, h) \in G_1^2$.

Definition 1. A bilinear parameter generator $\mathscr{G}en$ is a probabilistic algorithm that takes a security parameter κ as input, and outputs a 5-tuple (q, P, G_1, G_2, e).

1.4.2 PKE with Keyword Search

The concept of public key encryption (PKE) with keyword search (PEKS) was proposed by Boneh et al. [35], which supports the keyword search on encrypted data.

Fig. 1.3 A PECSK scheme framework

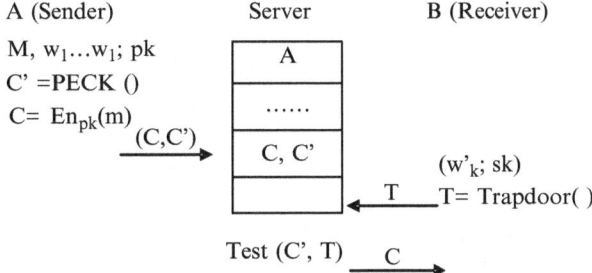

Formally, public key encryption with keyword search studies the problem of searching on data that is encrypted using a public key system. In this setting, the server plays the role of data warehouse for the receiver. The framework of PEKS scheme is illustrated in Fig. 1.3. With the PEKS scheme, the workflow of the underlying application consists of two phases.

1. In the first phase, a sender encrypts its message, runs PEKS() to generate some keyword tags for the message, and stores the ciphertext and the tags at the server.
2. In the second phase, the receiver runs Trapdoor() to generate a trapdoor for each selected keyword, and sends the trapdoors to the server, which will run Test() to search over the tags attached to each encrypted message.

Most of the existing literatures focus on the part of keyword search techniques PEKS(), they assume that any public key encryption can be used as En(). It does work in real applications, but if the foundation of En() and PEKS() are different, the overhead may be large. If there is a scheme which can combine the public key encryption En() and keyword search technique PEKS() together, it may improve the performance of the system. As such, it can be a real searchable encryption scheme.

1.4.3 HEV Based Query Predicate

The concept of Hidden Vector Encryption (HVE) is proposed by Boneh and Waters [40]. HVE is a type of predicate encryption where two vectors over attributes are associated with a ciphertext and a token, respectively. At a high level, the ciphertext matches the token if and only if the two vectors are component-wise equal. There are two character sets $\sum$ and $\sum_* = \sum \cup \{*\}$ in the setting of HVE. Here $\sum$ is an arbitrary set of attributes. We assume $\sum = \mathbf{Z}_q$; $*$ is a special symbol denoting a wildcard component, which means that the component related to $*$ is not involved with any attribute. HVE mainly consists of four phases: key generation, data encryption, token generation and data query. The notations used in the following will be described in Table 4.1.

1.4.3.1 Equality Query

- In key generation phase, the TA distributes the public/private key pair (PK, SK) to a receiver.
- In data encryption phase, a user chooses a vector $\mathbf{x} = (x_1, \ldots, x_l) \in \sum^l$ to characterize its data and encrypts its data m into a ciphertext CT using the receiver's public key.
- In token generation phase, the receiver chooses a vector $\mathbf{w} = (w_1, \ldots, w_l) \in (\sum_*)^l$ to represent his query requirements and generate a query token T_w. The receiver sends T_w to the server.
- In data query phase, if $\mathbf{x}$ equals to $\mathbf{w}$, the token can decrypt a ciphertext by using the receiver's private keys. The matching condition is defined as following: let $s(w)$ be the set of indexes i such that w_i is not a wildcard in the vector $\mathbf{w} = (w_1, \ldots, w_l)$. For the vector $\mathbf{x}$ and $\mathbf{w}$, let $P_\mathbf{w}(\mathbf{x})$ be the following equality predicate:

$$P_\mathbf{w}(\mathbf{x}) = \begin{cases} 1, \text{if for all } i \in s(w), w_i = x_i, \\ 0, \text{otherwise.} \end{cases} \tag{1.1}$$

Then, the server can disclose the data m if the equality predicate $P_\mathbf{w}(\mathbf{x}) = 1$.

1.4.3.2 Comparison Query

If we map the ith component $x_i \in \mathbf{x}$ to its domain $\{1, \ldots, n\}$ as in [43], the value of x_i is one of the number $j \in \{1, \ldots, n\}$. The key generation phase is same as in the above quality query.

Then, in the data encryption phase, the user builds an encryption vector $\sigma(\mathbf{x}) = (\sigma_{i,j}) \in \{0, 1\}^{nl}$ for $\mathbf{x} = (x_1, \ldots, x_l) \in \{1, \ldots, n\}^l$, as follows:

$$\sigma_{i,j} = \begin{cases} 1, \text{if } x_i \geq j, \\ 0, \text{otherwise,} \end{cases} \tag{1.2}$$

where, $i \in \{1, \ldots, l\}$ and $j \in \{1, \ldots, n\}$. For example, $l = 3, n = 5$ and let $\mathbf{x} = (1, 3, 2)$. Thus $\mathbf{x} = (x_1, \ldots, x_l) \in \{1, \ldots, n\}^l = \{1, 2, 3, 4, 5\}^3$ and the corresponding encryption vector $\sigma(\mathbf{x}) = (10000, 11100, 11000)$. Then, the data should be encrypted under the encryption vector $\sigma(\mathbf{x})$.

Next, in the token generation phase, the user builds a query vector $\sigma^*(\mathbf{w}) = (\sigma_{i,j}^*) \in \{0, 1, *\}^{nl}$ for $\mathbf{w} = (w_1, \ldots, w_l) \in \{1, \ldots, n\}^l$ as follows:

$$\sigma_{i,j}^* = \begin{cases} 1, \text{if } w_i = j, \\ *, \text{otherwise.} \end{cases} \tag{1.3}$$

Similarly, we assume $l = 3, n = 5$ and $\mathbf{w} = (w_1, \ldots, w_l) \in \{1, \ldots, n\}^l = \{1, 2, 3, 4, 5\}^3$. If the receiver's query condition is $P = (x_1 \geq 1) \wedge (x_2 \geq 3) \wedge (x_3 \geq 1)$, i.e., $\mathbf{w} = (1, 3, 1)$. Thus the query vector $\sigma^*(\mathbf{w}) = (1 * * * *, * * 1 * *, 1 * * * *)$. Note that, the number of the elements in $\sigma^*(\mathbf{w})$ is nl.

In the data query phase, let $s(\sigma^*(w))$ denotes the set of all indexes k which satisfies $\sigma_k^* \neq *$, where $k \in \{1, \ldots, nl\}$. Let $P_{\sigma^*(\mathbf{w})}(\sigma(\mathbf{x}))$ be the following comparison predicate:

$$P_{\sigma^*(\mathbf{w})}(\sigma(\mathbf{x})) = \begin{cases} 1, \text{if for all } i \in s(\sigma^*(w)), \sigma^*(w_i) = \sigma(x_i), \\ 0, \text{otherwise.} \end{cases} \qquad (1.4)$$

Finally, the server can disclose the data m if the comparison predicate $P_{\sigma^*(\mathbf{w})}(\sigma(\mathbf{x})) = 1$.

References

1. R. Lu, X. Liang, X. Li, X. Lin, and X. Shen, "Eppa: An efficient and privacy-preserving aggregation scheme for secure smart grid communications," *IEEE Transactions on Parallel and Distributed Systems*, vol. 23, no. 9, pp. 1621–1631, 2012.
2. G. FitzPatrick and D. Wollman, "Nist interoperability framework and action plans," in *Proc. PESGM*, pp. 1–4, IEEE, 2010.
3. W. Wang and Z. Lu, "Cyber security in the smart grid: Survey and challenges," *Computer Networks*, vol. 57, no. 2103, p. 1344–1371, 2013.
4. R. Sushmita and N. Amiya, "A decentralized security framework for data aggregation and access control in smart grids," *IEEE Transactions on Smart Grid*, vol. 4, no. 1, pp. 196–205, 2013.
5. M. M. Fouda, Z. M. Fadlullah, N. Kato, R. Lu, and X. Shen, "A lightweight message authentication scheme for smart grid communications," *IEEE Transactions on Smart Grid*, vol. 2, no. 4, pp. 675–685, 2011.
6. A. Hahn, A. Ashok, S. Sridhar, and M. Govindarasu, "Cyber-physical security testbeds: Architecture, application, and evaluation for smart grid," *IEEE Transactions on Smart Grid*, vol. 4, no. 2, pp. 847–855, 2013.
7. T. Liu, Y. Liu, and M. Yashan, "A dynamic secret-based encryption scheme for smart grid wireless communication," *IEEE Transactions on Smart Grid*, vol. pp, no. 99, pp. 1–8, 2013.
8. P. Jokar, N. Arianpoo, and V. Leung, "A survey on security issues in smart grids," *Security and Communication Networks*, 2012.
9. P. McDaniel and S. McLaughlin, "Security and privacy challenges in the smart grid," *IEEE Security & Privacy*, vol. 7, no. 3, pp. 75–77, 2009.
10. D. Von Dollen, "Report to nist on the smart grid interoperability standards roadmap," 2009.
11. The Smart Grid Interoperability Panel-Cyber Security Working Group, "Nistir 7628 guidelines for smart grid cyber security: Smart grid cyber security strategy, architecture, and high-level requirements." http://csrc.nist.gov/publications/nistir/ir7628/nistir-7628_vol1.pdf, August 2010.
12. A. Giani, E. Bitar, M. Garcia, M. McQueen, P. Khargonekar, and K. Poolla, "Smart grid data integrity attacks: characterizations and countermeasures π," in *Proc. SmartGridComm*, pp. 232–237, IEEE, 2011.

13. A. Giani, E. Bitar, M. Garcia, M. McQueen, P. Khargonekar, and K. Poolla, "Smart grid data integrity attacks," *IEEE Transactions on Smart Grid*, vol. 4, no. 3, pp. 1244–1253, 2013.
14. Y. Liu, P. Ning, and M. K. Reiter, "False data injection attacks against state estimation in electric power grids," *ACM Transactions on Information and System Security*, vol. 14, no. 1, p. 13, 2011.
15. G. Dán and H. Sandberg, "Stealth attacks and protection schemes for state estimators in power systems," in *Proc. SmartGridComm*, pp. 214–219, IEEE, 2010.
16. Y. Huang, M. Esmalifalak, H. Nguyen, R. Zheng, Z. Han, H. Li, and L. Song, "Bad data injection in smart grid: attack and defense mechanisms," *IEEE Communications Magazine*, vol. 51, no. 1, pp. 27–33, 2013.
17. X. Li, X. Liang, R. Lu, X. Shen, X. Lin, and H. Zhu, "Securing smart grid: cyber attacks, countermeasures, and challenges," *IEEE Communications Magazine*, vol. 50, no. 8, pp. 38–45, 2012.
18. H. Li and Z. Han, "Manipulating the electricity power market via jamming the price signaling in smart grid," in *Proc. Globecom Workshop*, pp. 1168–1172, IEEE, 2011.
19. A. R. Metke and R. L. Ekl, "Security technology for smart grid networks," *IEEE Transactions on Smart Grid*, vol. 1, no. 1, pp. 99–107, 2010.
20. M. Blaze, J. Feigenbaum, and A. Keromytis, "Keynote: Trust management for public-key infrastructures," in *Proc. SP*, pp. 59–63, Springer, 1999.
21. R. Ma, H.-H. Chen, Y.-R. Huang, and W. Meng, "Smart grid communication: Its challenges and opportunities," *IEEE Transactions on Smart Grid*, vol. 4, no. 1, pp. 36–46, 2013.
22. H. Khurana, R. Bobba, T. Yardley, P. Agarwal, and E. Heine, "Design principles for power grid cyber-infrastructure authentication protocols," in *Proc. HICSS*, pp. 1–10, IEEE, 2010.
23. Q. Li and G. Cao, "Multicast authentication in the smart grid with one-time signature," *IEEE Transactions on Smart Grid*, vol. 2, no. 4, pp. 686–696, 2011.
24. M. Kgwadi and T. Kunz, "Securing rds broadcast messages for smart grid applications," *International Journal of Autonomous and Adaptive Communications Systems*, vol. 4, no. 4, pp. 412–426, 2011.
25. S. Ruj and A. Nayak, "A decentralized security framework for data aggregation and access control in smart grids," *IEEE Transactions on Smart Grid*, vol. 4, no. 1, pp. 196–205, 2013.
26. V. Goyal, O. Pandey, A. Sahai, and B. Waters, "Attribute-based encryption for fine-grained access control of encrypted data," in *Proc. CCS*, pp. 89–98, ACM, 2006.
27. J. Bethencourt, A. Sahai, and B. Waters, "Ciphertext-policy attribute-based encryption," in *Proc. SP*, pp. 321–334, IEEE, 2007.
28. S. Ruj, A. Nayak, and I. Stojmenovic, "A security architecture for data aggregation and access control in smart grids," 2011.
29. A. Bartoli, J. Herna?ndez-Serrano, M. Soriano, M. Dohler, A. Kountouris, and D. Barthel, "Secure lossless aggregation for smart grid m2m networks," in *Proc. SmartGridComm*, pp. 333–338, IEEE, 2010.
30. F. Li, B. Luo, and P. Liu, "Secure information aggregation for smart grids using homomorphic encryption," in *Proc. SmartGridComm*, pp. 327–332, IEEE, 2010.
31. X. Liang, X. Li, R. Lu, X. Lin, and X. Shen, "Udp: Usage-based dynamic pricing with privacy preservation for smart grid.," *IEEE Transactions on Smart Grid*, vol. 4, no. 1, pp. 141–150, 2013.
32. B. Hore, S. Mehrotra, M. Canim, and M. Kantarcioglu, "Secure multidimensional range queries over outsourced data," *The International Journal on Very Large Data Bases*, vol. 21, no. 3, pp. 333–358, 2012.
33. R. Agrawal, J. Kiernan, R. Srikant, and Y. Xu, "Order preserving encryption for numeric data," in *Proc. SIGMOD*, pp. 563–574, ACM, 2004.
34. A. Boldyreva, N. Chenette, Y. Lee, and A. O'Neill, "Order-preserving symmetric encryption," in *Proc. EUROCRYPT*, pp. 224–241, Springer, 2009.
35. D. Boneh, G. Di Crescenzo, R. Ostrovsky, and G. Persiano, "Public key encryption with keyword search," in *Proc. Eurocrypt*, pp. 506–522, Springer, 2004.

36. P. Golle, J. Staddon, and B. Waters, "Secure conjunctive keyword search over encrypted data," in *Proc. ACNS*, pp. 31–45, Springer, 2004.
37. E. Shi, J. Bethencourt, T. Chan, D. Song, and A. Perrig, "Multi-dimensional range query over encrypted data," in *Proc. SP*, pp. 350–364, 2007.
38. J. Li and E. R. Omiecinski, "Efficiency and security trade-off in supporting range queries on encrypted databases," in *Proc. CDAS*, pp. 69–83, Springer, 2005.
39. J. Xu, J. Fan, M. H. Ammar, and S. B. Moon, "Prefix-preserving ip address anonymization: Measurement-based security evaluation and a new cryptography-based scheme," in *Proc. ICNP*, pp. 280–289, IEEE, 2002.
40. D. Boneh and B. Waters, "Conjunctive, subset, and range queries on encrypted data," in *Proc. TCC*, pp. 535–554, 2007.
41. S. De Capitani di Vimercati, S. Foresti, S. Paraboschi, G. Pelosi, and P. Samarati, "Efficient and private access to outsourced data," in *Proc. ICDCS*, pp. 710–719, IEEE, 2011.
42. K. LeFevre, D. J. DeWitt, and R. Ramakrishnan, "Mondrian multidimensional k-anonymity," in *Proc. ICDE*, pp. 25–25, IEEE, 2006.
43. J. Park, "Efficient hidden vector encryption for conjunctive queries on encrypted data," *IEEE Transactions on Knowledge and Data Engineering*, vol. 23, no. 10, pp. 1483–1497, 2011.

Chapter 2
Equality Query for Auction in Emerging Smart Grid Marketing

Distributed energy resources (DERs), which are characterized by small scale power generation technologies to provide an enhancement of the traditional power system, have been strongly encouraged to be integrated into the smart grid, and numerous trading strategies have recently been proposed to support the energy auction in the emerging smart grid marketing. However, few of them consider the security aspects of energy trading, such as privacy-preservation, bid integrity and pre-filtering ability. In this chapter, we introduce an efficient Searchable Encryption Scheme for Auction (SESA) in emerging smart grid marketing. Specifically, SESA uses a public key encryption with keyword search technique to enable the energy sellers, e.g. DERs, to inquire suitable bids while preserving the privacy of the energy buyers (EBs). Additionally, to facilitate the seller to search for detailed information of the bids, we also propose an extension of SESA to support conjunctive keywords search.

2.1 Introduction

Growing demand for electricity, upcoming fossil-fuel shortage and CO_2 emission crises have recently invoked an urgent need in incorporating renewable energy sources into the power grid. Such a trend is commonly known as distributed generation (DG) [1]. In the trend of DG, distributed energy resources have been encouraged to participate in energy marketing to facilitate competition among different energy providers. However, how to negotiate with different energy providers and energy consumers is a challenging issue in DG [2]. In order to address this challenge, smart grid, which is composed of many entities: intelligent electricity distribution devices, advanced sensors, two-way automated metering infrastructure, and specialized computer systems to enhance the operation performance [2], has received significant attention in recent years. Smart grid can accelerate the integration of distributed energy suppliers, DERs and microgrids [3], and thus it potentially makes power generation, transmission, and distribution be the next big e-business operating mostly under autonomous control [4]. As a result, smart grid has been recognized as an emerging electricity market.

In order to protect users' information privacy and security during the auction process, each buyer should protect their bidding information and let it not be known by the unauthorized users, including the auction server. While at the same time, it enables the sellers to query the auction server about the demanded bids. Although many auction models (e.g. [2, 5, 6]) were established respectively for smart grid energy marketing, few of them takes the privacy or security of the DERs into consideration. Recently, various security vulnerabilities and threats have been studied in the research literatures [7–9].

Lu et al. [10] used a super-increasing sequence to structure multi-dimensional data and encrypt the structured data by the homomorphic paillier cryptosystem technique. Li et al. [11] proposed an efficient demand response scheme to achieve privacy preserving demand aggregation and efficient response. However, since these encryption schemes can not be searched, they are not suitable for auction in smart grid marketing. On the other hand, some of the traditional auction schemes [12, 13] can achieve bidding privacy, but they can not support keyword search or bids filtering.

In this chapter, we address the efficient searchable encryption problem for auction in smart grid marketing. This scheme considers both the public key based encryption and keyword search techniques. It can achieve privacy-preservation, searchable ability and bids filtering, as well as other security features including confidentiality, authenticity and integrity. The main content of this chapter are twofold.

1. Firstly, a novel SESA scheme is constructed to achieve searchable encryption, by modifying the proxy re-encryption with keyword search scheme [14]. The security analysis demonstrates that SESA can achieve confidentiality, data and keyword privacy, authenticity and data integrity.
2. Secondly, we construct an extended version of SESA to support conjunctive keywords search. It enables the user to question the auction server more flexibly.

The remainder of this chapter is organized as follows. In Sect. 2.2 we describe the smart grid marketing architecture, security requirement and design goal. Then, we present the SESA scheme and its extension in Sects. 2.3 and 2.4, followed by its security analysis and performance evaluation in Sects. 2.5 and 2.6, respectively. Next, we review the related works in Sect. 2.7. Finally, we draw our summary in Sect. 2.8.

2.2 System Model and Design Goal

In this section, we formalize the system model, security requirements, and identify our design goals.

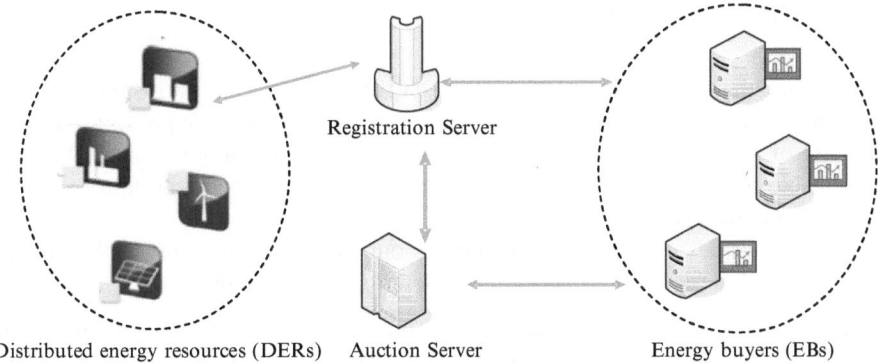

Distributed energy resources (DERs)   Auction Server   Energy buyers (EBs)

Fig. 2.1 Smart grid marketing architecture

2.2.1   Smart Grid Marketing Architecture

Smart grid marketing refers to a system that enables small producers to generate and sell electricity at the local level. As shown in Fig. 2.1, there are energy sellers (e.g. DERs), energy buyers (e.g. EBs), and auction managers. The auction managers are two servers: a registration server (RS) and an auction server (AS).

RS: In an energy marketing, a registration server is used to initiate the system at the beginning of the auction; and when the bidding is finished, it will select the winner according to the criteria of the DERs. The RS is trustworthy and it will send some keywords from the DERs to the auction server to search for their wanted bids. The winner may be selected from these pre-filtered bids.

AS: Auction server is used in a continuous sealed-bid auction in which traders submit offers to buy (bid) or offers to sell (ask) at any time during the trading period. The auction server is semitrust and it cannot know the content of the EBs' bids, but it can test if the message has tags like the seller's query.

DER: DERs can open the bids by themselves. However, due to the number of distributed bids from EBs may be large, to improve the efficiency, the RS will act as a proxy for the DERs to select the winners.

EB: Energy is bought from or sold to the grid depending on the availability, demand, and price of energy. Each energy buyer will send its sealed-bid to the auction server. Due to the large amount of buyers, the bids may be conducted with the competition of others.

2.2.2 Security Requirements

We assume that the communication between EBs and server is untrustworthy. That is, various adversaries such as eavesdroppers and tampers may be present. If a large amount of EBs are competitive to buy a certain type of energy from DERs, it is reasonable to enable the RS to query the AS and select one or group of winners according to the criteria of the DERs.

We define the security requirements for our SESA scheme, and will show the fulfillment of these requirements after presenting the design details.

- *Data privacy*: The data owner can resort to the public key cryptography to encrypt the data before outsourcing, and successfully prevent the unauthorized entities, including the auction server, from prying into the outsourced data.
- *Bid integrity*: The bids information and queries should not be changed by the malicious users or the illegal competitors, i.e., if the competitor $\mathscr{A}$ maliciously modified the price or other information of EB_i, it may lead to EB_i can not be selected by the RS.
- *Keyword privacy*: As users usually prefer to keep their search from being exposed to others, including the auction server, the most important concern is to hide what in their bids and what the RS is inquiring, i.e., the keywords indicated by the corresponding trapdoor. Thus, the trapdoor should be generated in a cryptographic way to protect the query keywords.
- *Trapdoor unforgeability*: DER generates his trapdoor information based on his keyword and secret key. After the AS receiving the trapdoor, it can test this trapdoor with keyword tags. The most important thing is that others (include the AS) can get nothing from the trapdoor, i.e. the AS cannot forge a new trapdoor based on the old ones.

2.2.3 Design Goal

To enable searchable encryption for effective utilization of outsourced energy bids under the aforementioned model, our design goal is to develop a searchable encryption scheme for auction in emerging smart grid marketing, and achieve the security of the bids and efficient keyword search as follows.

- The proposed scheme should achieve security as mentioned in the security requirements, i.e., the data privacy, keyword privacy, data integrity and trapdoor unforgeability.
- The proposed scheme should achieve both one keyword and conjunctive keywords search.
- The proposed scheme should achieve the communication and computation efficiency, compared with other searchable encryption schemes.

2.3 SESA Scheme

In this section, we show the construction of the efficient Searchable Encryption Scheme for Auction (SESA) in emerging smart grid marketing, which mainly consists of the following four phases: Registration phase, Bidding phase, Pre-filtering phase, and Decision-of-winner phase. For our auction system, we assume that there is a local registration server (RS) which can bootstrap the system. Specifically, in this system initialization phase, given the security parameter 1^k, RS first generates (q, g, G_1, G_2, e) by running $\mathscr{G}en(1^k)$, where q is a k-bit prime number. Let G_1 and G_2 be two cyclic multiplication groups. $Sig(\mathscr{G}, U, V)$ is an identity based signature scheme [15]. Furthermore, we will need three hash functions $H_1 : \{0, 1\}^* \rightarrow G_1$, $H_2 : \{0, 1\}^* \rightarrow G_1$, $H_3 : G_2 \rightarrow \{0, 1\}^*$. RS publishes the system parameters as $(q, g, G_1, G_2, e, H_1, H_2, H_3)$.

2.3.1 Registration Phase

In order to maintain security of the network against attacks and the fairness among customers and providers, the local RS may control the access of each DER and EB. The energy marketing announces two prices: the price for selling energy and the price for buying energy in the smart grid marketing. The DERs adjust their bidding price after negotiating with the other units based on the grid prices, considering their operational cost and local demands.

In our scheme, there are n DERs and m EBs in the energy marketing. For each DER_i (i $= 1, \ldots, n$) and EB_j (j $= 1, \ldots, m$), when they register, the RS picks two random numbers $x_i, y_i \in Z_q^*$ and sets $pd_i = g^{x_i}$, $pb_j = g^{y_j}$. (pd_i, x_i) and (pb_j, y_j) are DER_is and EB_js public/private key pairs, respectively. For each EB_j (j $= 1, \ldots, m$), the RS randomly chooses a master key $s \in Z_q^*$ and assigns an ID-based key pair $(H_1(ID_{EB_j}), H_1^s(ID_{EB_j}))$ to DER_i for signature, and denotes it as (vk_j, ssk_j).

In the energy marketing, the DER_i will publish its energy information $m_i = (p_i, GID_i, Ts, Lo_i, Am_i, T_N)$ publicly, where p_i is the initial price, GID_i is the identification of the energy, Ts is the timestamp, Lo_i is the energy resource location, Am_i is the amount of the energy and T_N is the unique serial number of the deposit energy information. The RS will store the information from each DER_i as a tuple (DER_i, m_i) in its database. Also, EB_j will register its personal information $e_j = (Lo_j, Rep_j, Ty_j, \Delta)$ on the RS, where Lo_j is its location, Rep_j is its reputation about its history trades (which also will be verified by the RS, but it's not our paper's focus), Ty_j is the demanded energy types, Δ is the other information of EB_j. The RS also stores the information from each EB_j as a tuple (EB_j, e_j) in its database.

2.3.2 Bidding Phase

In order to achieve the nearly real-time energy bidding, each EB_j will choose its interested energy to bid. The bidding is performed as following steps.

1. EB_j gets an identity based signature key pair as (vk_j, ssk_j) from RS. The public key is represented as $A = vk_j$, and the private key ssk_j is kept secretly.
2. EB_j selects a random $r_j \in Z_q^*$, and generates a $bid_j = (EB_j, pr_j, GID_i, Cb_j, Rep_j)$, where pr_j is the price of the bid, Cb_j is the amount of the energy that EB_j wants to buy, Rep_j is EB_j's reputation. Then, EB_j computes $C_j = H_3(e(g, H_2(A)^{r_j})) \oplus bid_j$.
3. In order to maximize the probability of winning in the auction, EB_j selects a keyword w_j to represent his bid (e.g.the reputation or required amount). Next, EB_j computes a tag on the keyword as $t_j = e(g, H_1(w_j)^{r_j})$. Then he computes $B_j = (g^{x_i})^{r_j}$ and $F_j = H_3(t_j)$. He outputs $C_j' = (B_j, F_j)$.
4. EB_j generates a signature $S_j = Sig_{ssk_j}(C_j, C_j')$. (C_j, C_j') is the signed message.
5. EB_j sends the encrypted message $K_j = (GID_i, A, C_j, C_j', S_j)$ to the auction server.
6. The auction server stores this information from EB_j as a tuple (EB_j, K_j) in its bid table.

2.3.3 Pre-filtering Phase

The goal of bids pre-filtering is to quickly identify potential winner or winners from all the bids in the AS's bid table.

For example, if DER_i wants to filter the bids for energy GID_i according to the user's reputation w_i', DER_i generates a trapdoor $t_{w_i'}$ in advance and sends it to the RS. In order to preserve the privacy of DER_i and EB_j, the trapdoor $t_{w_i'} = H_1(w_i')^{1/(x_i)}$ is a ciphertext of the value w_i'. Then RS will send $t_{w_i'}$ to the AS. On receiving the message from RS, for each bid in the AS's bid table, the auction server will test if the given C_j' satisfies the selection criterion $t_{w_i'}$ of DER_i:

1. Message verification:

   (a) The auction server verifies signature S_j on message (C_j, C_j') with respect to the public key A.
   (b) If it fails, the auction server will reject this bid; else the auction server will go on testing.

2. Trapdoor and tag test:
   The auction server tests if $H_3(e(B_j, t_{w_i'})) = F_j$. If so, which means $w_j = w_i'$; and the encrypted bid C_j will be stored in a filtered array W[]. Later, W[] will be transferred to the RS. If not, AS will go on testing the other bids. The correctness of $H_3(e(B_j, t_{w_i'})) = F_j$ is as follows:

$$H_3(e(B_j, t_{w_i'})) = H_3(e((g^{x_i})^{r_j}, H_1(w_i')^{1/(x_i)}))$$

$$= H_3(e(g, H_1(w_i')^{r_j}))$$

$$= F_j = H_3(e(g, H_1(w_j)^{r_j}))$$

2.3.4 Decision-of-Winner Phase

On receiving filtered bids array $\mathbf{W}[]$ from the AS, the RS can decrypt each C_j in $\mathbf{W}[]$ as $bid_j = C_j \oplus H_3(e(B_j, H_2(A))^{1/x_i})$ by using DER_i' secret key x_i, otherwise, C_j will be discarded. The correctness of the decryption is shown as follows,

$$C_j \oplus H_3(e(B_j, H_2(A))^{1/x_i}) = H_3(e(g, H_2(A)^{r_j})) \oplus bid_j \oplus H_3(e((g^{x_i})^{r_j}, H_2(A))^{1/x_i})$$

$$= H_3(e(g, H_2(A)^{r_j})) \oplus bid_j \oplus H_3(e(g, H_2(A)^{r_j})) = bid_j$$

We assume that there are t decrypted bids, and the bids will be put in a sorted array list $\mathbf{B}[]$ according to their price descending order. Due to the special difficulties in energy storage and profit maximization of the auction in nature, the winner-selection criterion from DER_i should achieve two goals: one is that the total sales should be as high as possible; the other is that the sum of the demanded amount of the winners should be as close to the available energy demand Am_i as possible. The selected winners will be stored in an array list $\mathbf{S}[]$ by using Algorithm 1. Finally, the RS will secretly deliver the winners list $\mathbf{S}[]$ to DER_i.

2.4 Extended SESA with Conjunctive Keywords Search

The SESA can be extended to support conjunctive keywords search, with which the DERs can get more detailed information about the bids. Since the Decision-of-winner phase in this extension is same as that in SESA, we only introduce the Registration phase, Bidding phase and Pre-filtering phase as following.

2.4.1 Registration Phase

In this extended scheme there are also n DERs and m EBs in the energy marketing. For each DER_i ($i = 1, \ldots, n$) and EB_j ($j = 1, \ldots, m$), when they register, the RS picks two random numbers $x_i, y_i \in Z_q^*$ and sets $pd_i = g^{x_i}$, $pb_j = g^{y_j}$. (pd_i, x_i) and (pb_j, y_j) are DER_is and EB_js public/private key pairs respectively.

Algorithm 1 Winner selection(**B,S**)

1: $c \leftarrow Am_i$, c is the remain energy amount;
2: $k1, k2 \leftarrow 0; k = t; S[\ ] \leftarrow \phi$.
3: If two bids have the same price, the one requires bigger amount will be first served.
4: **while** $k \neq 0$ **do**
5:     **for** each **B**[$k1$] **do**
6:        **if** (**B**[$k1$].$price$ = **B**[$k1 + 1$].$price$) and (**B**[$k1$].$amount$ < **B**[$k1 + 1$].$amount$) **then**
7:           $temp \leftarrow$ **B**[$k1$];
8:           **B**[$k1$] $\leftarrow$ **B**[$k1 + 1$];
9:           **B**[$k1 + 1$] $\leftarrow temp$;
10:       **end if**
11:       $k1 + +$;
12:    **end for**
13:    $k - -$;
14: **end while**
15: $k1 \leftarrow 0$;
16: **for** each **B**[$k1$] **do**
17:    **if** (**B**[$k1$].$amount$ < c) **then**
18:       $S[k2] \leftarrow$ **B**[$k1$], $k2 + +$;
19:       $c \leftarrow c -$ **B**[$k1$].$amount$;
20:    **end if**
21:    $k1 + +$;
22: **end for**
23: **return** (**S**[]);

The RS randomly chooses a master key $s \in Z_q^*$ and assigns an ID-based key pair $(H_1(ID_{EB_j}), H_1^s(ID_{EB_j}))$ for each EB_j (j $= 1, \ldots, $m). The key pair is represented as (vk_j, ssk_j). Similar to the SESA, the DER_i will publish its energy information $m_i = (p_i, GID_i, Ts, Lo_i, Am_i, T_N)$ publicly. The RS will store the information from each DER_i as a tuple (DER_i, m_i) in its database. Also, EB_j will register its personal information $e_j = (Lo_j, Rep_j, Ty_j, \triangle)$ on the RS. The RS also stores the information from each EB_j as a tuple (EB_j, e_i) in its database.

In order to provide more convenience for the DERs to get detailed filtering, i.e. let them achieve the conjunctive keywords search from the auction server, each EB_j will select a keywords set $W_j = \{w_{j1}, w_{j2}, \ldots, w_{jL}\}$ to characterize his bid. Without loss of generality, the location of each type of keyword in the keywords set $W_j = \{w_{j1}, w_{j2}, \ldots, w_{jL}\}$ is fixed. For instance, w_1 denotes the type of the source address keyword, w_2 denotes the type of energy amount keyword etc. Keywords in the DER_i' tag and EB_j' trapdoor are in the same order.

2.4.2 Information Encryption

Each EB_j publishes its bid as following steps:

1. EB_j gets an identity based signature key pair as (vk_j, ssk_j). The public key is denoted as $A = vk_j$, and the private key ssk_j is kept secretly.

2. EB_j selects a random number $r_j \in Z_q^*$, and generates a $bid_j = (EB_j, pr_j, GID_i, Cb_j, Ts_j, \triangle)$, where pr_j is the price of the bid, Cb_j is the amount of the energy that EB_j want to buy, $\triangle$ is the other information of EB_j. Then EB_j computes $C_j = H_3(e(g, H_2(A)^{r_j})) \oplus bid_j$.

3. EB_j computes a tag for each keyword as $t_{jk} = e(g, H_1(w_{jk})^{r_j}), (k = 1, \ldots, L), B_j = (g^{x_i})^{r_j}$. EB_j outputs $C_j' = (B_j, t_{jk}(k = 1, \ldots, L))$.

4. EB_j generates a signature $S_j = S_{ssk}(C_j, C_j')$, where the message to be signed is the tuple (C_j, C_j').

5. EB_j sends the encrypted messages $K_j = (A, C_j, S_j, C_j')$ to the auction server.

6. The auction server will store this information from EB_j as a tuple (EB_j, K_j) in its bid table.

2.4.3   Pre-filtering Phase

If the DER_i needs to filter the bids by using some criteria (e.g. reputation, location etc.). It will generate a keywords set $Q_i = \{w_{E1}, w_{E2}, \ldots, w_{Et}\}$. Then DER_i generates a trapdoor t_{Q_i} and sends it to the RS. At the end of the auction, the RS will transfer this trapdoor t_{Q_i} to the AS to filter the bids. Without loss of generality, we assume, $\{E1, E2, \ldots, Et\}$ is the subset of $\{j1, j2, \ldots, jL\}$.

1. DER_i generates a trapdoor on the keywords Q_i as $t_{Q_i} = (H_1(w_{E1}).H_1(w_{E2}) \ldots H_1(w_{Et}))^{1/(x_i)}$. DER_i sends $(t_{Q_i}, \{E1, E2, \ldots, Et\})$ to the RS. The RS transfers them to the AS.

2. For each C_j in GID_is bid table, the auction server will test if C_j' satisfies the EB_js requirement:

(1) Message verification:

   (a) The auction server verifies signature S_j on message (C_j, C_j') with respect to the public key A.

   (b) If it fails, the auction server will reject this bid; else the auction server will go on testing.

(2) The AS tests if $H_3(e(B_j, t_{Q_i})) = H_3(\prod_{v=E1}^{Et} t_v)$. If so, C_j will be stored in an array list W[]; if not, C_j will be rejected. The correctness of the test is shown as follows:

$$H_3(e(B_j, t_{Q_i})) = H_3(e((g^{x_i})^{r_j}, (H_1(w_{E1}).H_1(w_{E2}) \ldots H_1(w_{Ek}))^{1/(x_i)}))$$

$$= H_3(e(g, H_1(w_{E1})^{r_j}).e(g, H_1(w_{E2})^{r_j}) \ldots e(g, H_1(w_{Ek})^{r_j})) = H_3(\prod_{v=E1}^{Ek} t_v)$$

2.5 Security Analysis

In this subsection, we analyze the security properties of the proposed SESA scheme. In particular, following the security requirements discussed earlier, our analysis will focus on how the proposed SESA scheme can achieve the goals. The extension can also achieve these properties.

- *The individual EB's bid is privacy-preserving in the proposed SESA*: In the proposed SESA scheme, EB's bidding information is encrypted by its secret number r_j as $C_j = H_3(e(g, H_2(A)^{r_j})) \oplus bid_j$. Anyone, including the auction server who does not know the secret number r_j can not recover bid_j from the ciphertext C_j. Thus, if a bidder does not win the auction, in the proposed SESA nobody can get any information about the bidder from its bid.
- *The authentication and data integrity of the individual EB's bid is achieved in the proposed SESA*: In SESA, each EB's bidding information is signed by the identity based signature scheme [15]. Since the identity based signature $S_j = S_{ssk}(C_j, C'_j)$ is provably secure, the source authentication and data integrity can be guaranteed. As a result, the adversary $\mathscr{A}$'s malicious behaviors in the smart grid communications can be detected in the proposed SESA.
- *The EB's keyword privacy and DER's trapdoor privacy are also achieved in the proposed SESA*: In the proposed SESA, on one hand, the keyword which EB chose to append on the encrypted bid is protected by a hash function. Anyone, including the AS, can not recover w_j with the message C'_j. On the other hand, when RS delivers DER's query to the AS to search for certain type of bids, the query is also not delivered by plaintext, it is protected by a hash function. Thus, anyone gets the trapdoor only know the hash value of the keyword w'_i, and they do not know what the DER is really inquiring. Even when the AS does the verification of the tag and the trapdoor, it can not know anything about the keyword except for whether the they match or not.
- *The DER's trapdoor can not be forged in the proposed SESA*: In the proposed SESA, although the AS can get lots of trapdoors from DERs, it can not forge a valid new one from the existing old ones. That is because all the keywords are blinded by a hash function, the AS can not get the real value of the keywords.

Table 2.1 Comparison of security properties

Properties	Scheme [6, 16]	Scheme [12]	Scheme [13]	SESA [17]
Confidentiality	No	Yes	Yes	Yes
Data privacy	No	No	Yes	Yes
Bid integrity	No	Yes	Yes	Yes
Keyword privacy	No	No	No	Yes
Trapdoor unforgeability	No	No	No	Yes

It is illustrated in Table 2.1 that most of the auction schemes [6, 16] for power market are lack of security concerns. While in traditional electronic auction system, the work in [12] only achieves the confidentiality and data integrity, the work in

[13] achieves confidentiality, data privacy and data integrity. Only the proposed SESA scheme can achieve additional keyword privacy and trapdoor unforgeability compared with [13].

Figure 2.2 shows that if the auction server is compromised, the bids information and bidder's privacy will be disclosed in schemes [6, 13, 16], only those in [13] and the proposed SESA scheme can remain secure. But [13] can not support keyword search on the bids, and there is only one winner in [13]; it is not applicable for energy auction in the smart grid. From the above analysis, we can see the proposed SESA scheme can provide enough security guarantees for auction in smart grid marketing.

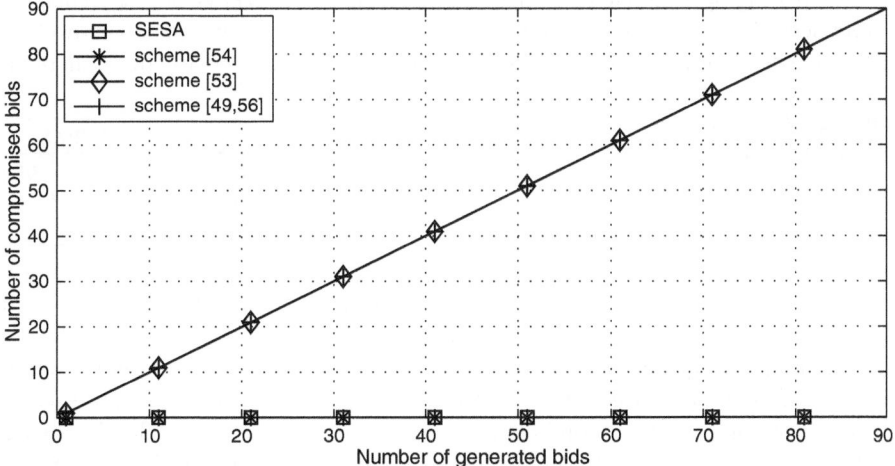

Fig. 2.2 Fraction of compromised bids when the auction server is compromised

2.6 Performance Analysis

2.6.1 SESA vs. EPPKS

In this subsection, we will compare our SESA with the privacy preserving keyword search scheme (EPPKS) [18] in terms of the computation and communication overhead in the one keyword search process.

Computation: In our proposed SESA, the computation tasks include pairing operations and exponentiation operations, where the pairing operations are the most time-consuming tasks. Since the hash operation and number multiplication are too fast compared with the pairing operations, we will not take them into consideration in this subsection. For simplicity of description, the pairing operation and exponentiation operation are denoted as C_p and C_e, respectively.

For the proposed SESA scheme, when an energy buyer EB_j generates an encrypted bid (A, C_j, S_j, C'_j), it requires 3 exponentiation operations and 2 pairing operations for bid encryption generation, i.e. $2C_p + 3C_e$. The DER_i or the RS needs 1 exponentiation operations to compute a trapdoor $t_{w'_i}$. After receiving the trapdoor from DER_i, the local AS needs to compute 2 pairings to verify the signature [15] and 1 pairing to test if there is a bid satisfying DER_i's query. Finally, DER_i or the RS requires 1 pairing operation and 1 exponentiation operation to decrypt the ciphertext if there are suitable bids.

In comparison, for EPPKS [18], it needs 3 pairing operations and 6 exponentiation operations to generate a data encryption on one keyword, i.e. $3C_p + 6C_e$. The seeker needs 1 exponentiation operation to compute a trapdoor T_{w_i}. And the server needs 1 pairing operation to test whether a given tag contains keyword T_{w_i}. Then the server needs $2C_p + 2C_e$ more computation overhead to get an intermediate result of the partial decipherment. At last, it will cost the seeker C_e to recovery the ciphertext.

Table 2.2 Comparison of computation complexity

	SESA	EPPKS
EB	$2C_p + 3C_e$	$3C_p + 6C_e$
AS	$3C_p$	$3C_p + 2C_e$
DER_i or RS	$C_p + 2C_e$	$2C_e$

Table 2.2 indicates that SESA is more efficient than EPPKS [18]. Detailed experiments also are conducted on a Pentium IV 3 GHz system to study the execution time [19]. For G_1 over the FST curve, a single exponentiation operation in G_1 with 161 bits costs 1.1 ms and the corresponding pairing operation costs 3.1 ms. The comparison of computation overhead is shown in Fig. 2.3. We can see that SESA achieves totaly lower execution times compared to EPPKS. Moreover, SESA can guarantee the integrity of the message, while EPPKS can not achieve this property.

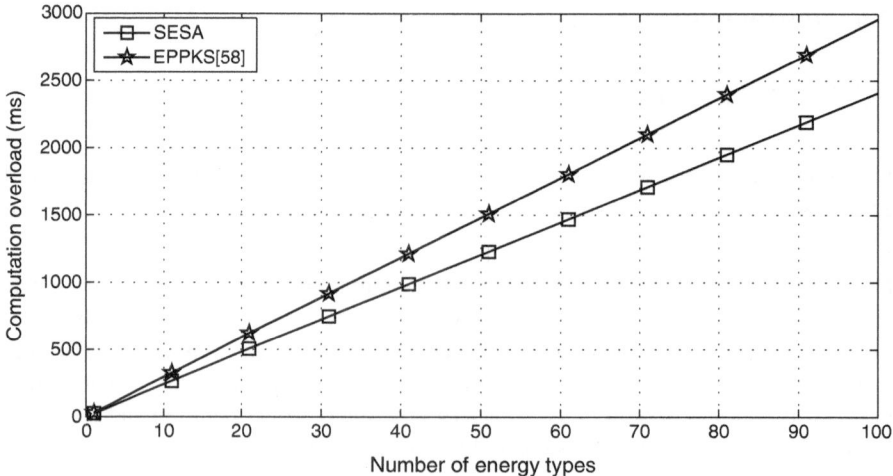

Fig. 2.3 Comparison of computation overhead between SESA and EPPKS

Communication: Most pairing-based cryptosystems need to work in a subgroup of the elliptic curve $E(F_q)$. By representing elliptic curve points using point compression [20], the length of the elements in G_1 and G_2 will be roughly 161-bit (using point compression) and 1,024-bit, respectively. SHA-1 is used to compute the hash function, which yields a 160-bit output. Let the parameter n in EPPKS be 160-bit. The communications among the three entities of the proposed SESA can be divided into three parts, EB-to-AS, DER-to-AS, and AS-to-RS communications.

We first consider the EB-to-AS communication in SESA. In the information encryption phase, the data report is in the form of $K_j = (A, C_j, S_j, C'_j)$. Since the length of identity based signature [15] is two G_1 elements, the size of K_j should be $160 + 160 + 161 * 2 + 160 + 161 = 963$ bits. In the DER-to-AS communication, DER needs to delivery a trapdoor t'_w to the AS, which is 160 bits; while in AS-to-RS communication, the AS will reply a ciphertext C_j to the EB if there is energy matching EB's demand, which is 160 bits.

Table 2.3 Comparison of communication complexity(bits)		SESA	EPPKS
	EB-to-AS	803	640
	DER-to-AS	160	160
	AS-to-RS	160	1,665

In contrast, the user-to-server communication overhead in EPPKS is the message (C_m, C_w), which includes one G_1 element, two n-bit elements and one hash element. The size is $161 + 2n + 160 = 641$ bits. Then, the trapdoor T_{w_j} with the size of 160-bit will be sent from user to the server. In the server-to-receiver communication, if there is a keyword match, the server will reply (C_m, C_ρ, C_w) to the receiver. Here, C_ρ is an element of G_2. The size of the reply is $161 + 160 + 2n + 1,024 = 1,665$ bits. Table 2.3 and Fig. 2.4 show the comparison of communication overhead between SESA and EPPKS. It can be seen that the SESA scheme significantly reduces the communication overhead.

2.6.2 Extended SESA vs. EPPKS

In this subsection, we will compare our extension of SESA with the privacy preserving keyword search scheme (EPPKS) [18] in terms of the computation overhead in the conjunctive keywords search process. Suppose there are 10 keywords tags on each bid, and 5 keywords in the EB'_j conjunctive search trapdoor. In the extension, it costs the EB_j $10 + 1$ pairing operations and $10 + 2$ exponentiation operations to generate an energy encryption (A, C_j, S_j, C'_j). That is $11C_p + 13C_e$. while the DER_i or the RS needs $5 + 2$ hash operations and 1 exponentiation operation to compute the trapdoor. On receiving the trapdoor t'_w from RS, the local AS needs 5 pairing operations and 1 hash operation to test DER_is query. If there is a suitable bid, the local AS needs to compute 2 pairings to verify the signature

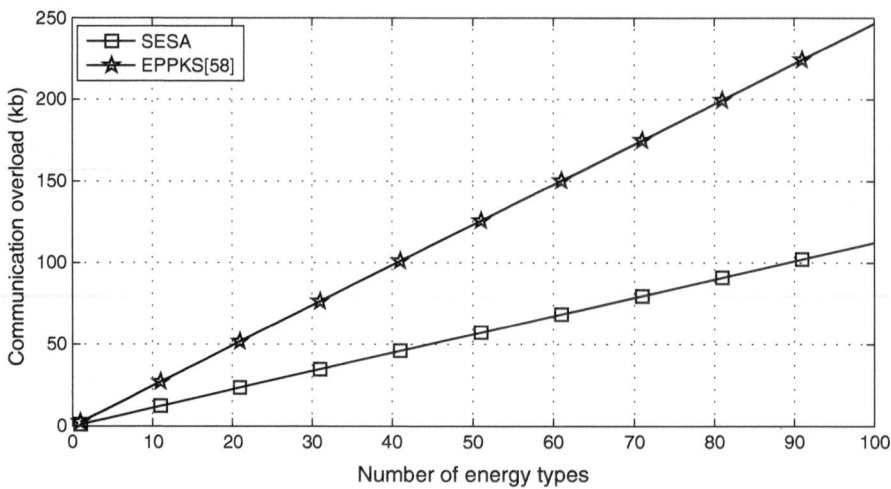

Fig. 2.4 Comparison of communication overhead between SESA and EPPKS

and 1 pairing to test if there is an energy satisfy DER_i's load demand. The DER_i or RS requires 1 pairing operation and 1 exponentiation operation to decrypt the ciphertext.

Table 2.4 Comparison of computation complexity

	SESA	EPPKS
EB	$11C_p + 12C_e$	$12C_p + 24C_e$
AS	$3C_p$	$7C_p + 2C_e$
DER_i or RS	$C_p + 2C_e$	$6C_e$

In comparison, the EPPKS needs $10 + 2$ pairing operations and $10 * 2 + 4$ exponentiation operations to generate an energy encryption on 10 keywords. That is totally $12C_p + 24C_e$. Since EPPKS can do 1 keyword search at a time, for 5-keyword search, the seeker needs to compute 5 trapdoors and sends them to the server, which needs 5 exponentiation operations. Thus, the server needs to test 5 times. Each time, the server needs 1 pairing operation to test whether a given tag contains keyword T_{w_i} or not. Thus, the server totaly needs $5C_p$ to test all of the trapdoors. If there is a matching item, the server needs $2C_p + 2C_e$ more computation overhead to get an intermediate result of the partial decipherment. At last, it will cost the seeker C_e to recovery the ciphertext.

From Table 2.4 and Fig. 2.5, it can be seen that the extension of SESA requires much less computation overhead than the EPPKS for the conjunctive keywords search. In addition, the extension is also more efficient than the EPPKS in terms of communication overhead, because more trapdoors need to be sent to the server in EPPKS.

2.7 Related Works

The traditional auction schemes can be divided into two categories: open outcry and sealed bids. Open outcry can further be separated into English auctions and Dutch auctions [16]. In English auctions, the value of the bid is public, and the price of the bid must be higher than the current price. The highest bidder is the winner at the end of the bidding phase. There are many famous English auction web sites (e.g., Yahoo!, eBay, etc.) [13]. The Dutch auction is almost the same as the English auction, except that it begins with the top price. In a sealed bid auction, the bidders write the price and quantity of their bid on a sheet of paper, and then they seal the sheet and give it to the auctioneer. The auctioneer collects all the sealed sheets and opens them after the deadline to determine the winner. A sealed bid auction can be separated into two kinds, first-price sealed-bid and second-price sealed-bid.

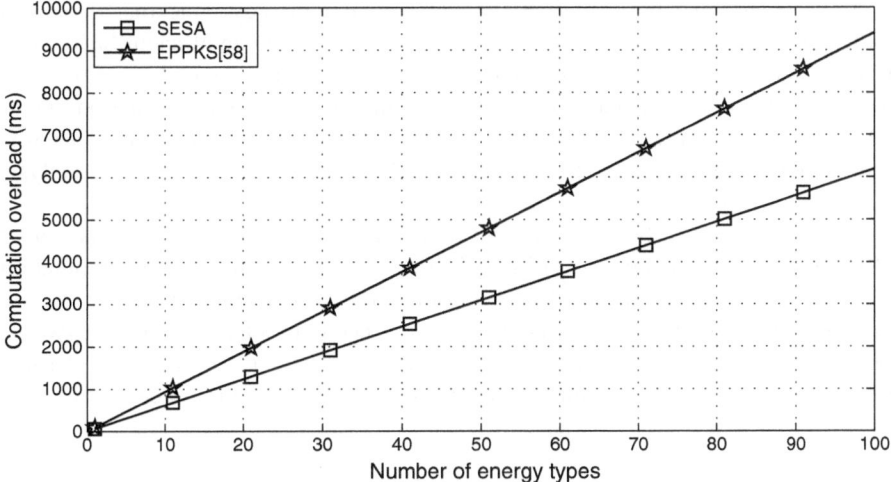

Fig. 2.5 Comparison of computation between extended SESA and EPPKS

The bidding manner has been extensively studied and various bidding models are presented in the power market [5,6]. Among the various methods, the simplest way is to estimate the market clearing price of the next time and then present the bid with a lower price than the estimated one. The second method is to estimate the behaviors of the rivals and to present the bid [6]. The third method is based on the game theory [21] with oligopolistic strategy such as Cournot model, and supply function models [5]. But, few of them considers the privacy of the bidders and the energy providers. In electronic auction systems, Chang [12] and Li [13] both presented anonymous auction protocol with freewheeling bids. However, bidding privacy can not be achieved in [12], and both of them can not support keyword search or any other filtering.

The concept of public key encryption with keyword search (PEKS) was proposed by Boneh et al. [22], which supports the keyword search on encrypted data. Other schemes focusing on constructing keyword encryption were extensively discussed, such as [23]. PECSK [24] supports conjunctive-subset keywords search. But it is only a keyword search scheme. EPPKS [18] presented a privacy preserving keyword search scheme in cloud computing. It is one of the few schemes which integrates both the message encryption and keyword search properties. However, when the server finds a tag matching the trapdoor in EPPKS [18], the server has to compute an intermediate result to help the user to recover the message, which costs communication and computation overhead.

2.8 Summary

In this chapter, we have studied the security and privacy concerns associated with energy auction in smart grid marketing, and proposed an efficient Searchable Encryption Scheme for Auction. We use public key encryption with keyword search to enable the energy sellers to inquire potential winner from the auction server while preserving the privacy of the EBs. In addition, an extension of SESA was presented to support detailed filtering of the bids. Security and performance analysis demonstrate that our proposed SESA and its extension both can achieve data and keyword privacy, bid integrity and trapdoor unforgeability, and they are more efficient than the existing keyword search approach EPPKS in terms of computation and communication overhead. However, for the multidimensional data in smart grid, in some cases, the conjunctive keyword query needs to support subset keywords query function for flexible usage. In the subsequent chapters, we will consider this problem and address the conjunctive query over encrypted multidimensional data.

References

1. C. Yuen, A. Oudalov, and A. Timbus, "The provision of frequency control reserves from multiple microgrids," *IEEE Transactions on Industrial Electronics*, vol. 58, no. 1, pp. 173–183, 2011.
2. B. Ramachandran, S. K. Srivastava, C. S. Edrington, and D. A. Cartes, "An intelligent auction scheme for smart grid market using a hybrid immune algorithm," *IEEE Transactions on Industrial Electronics*, vol. 58, no. 10, pp. 4603–4612, 2011.
3. V. Forte, "Smart grid at national grid," in *Proc. ISGT*, pp. 1–4, IEEE, 2010.
4. S. Chakraborty, M. D. Weiss, and M. G. Simões, "Distributed intelligent energy management system for a single-phase high-frequency ac microgrid," *IEEE Transactions on Industrial Electronics*, vol. 54, no. 1, pp. 97–109, 2007.
5. E. Bompard, W. Lu, and R. Napoli, "Network constraint impacts on the competitive electricity markets under supply-side strategic bidding," *IEEE Transactions on Power Systems*, vol. 21, no. 1, pp. 160–170, 2006.

6. Y.-q. SONG, L.-w. JIAO, Y.-x. NI, F.-s. WEN, Z.-j. HOU, and F.-l. WU, "An inproovement of generation firms' bidding strategies based on conjectural variation regulation via dynamic learning," *Proceedings of the Csee*, vol. 12, p. 004, 2003.

7. X. Li, X. Liang, R. Lu, X. Shen, X. Lin, and H. Zhu, "Securing smart grid: cyber attacks, countermeasures, and challenges," *IEEE Communications Magazine*, vol. 50, no. 8, pp. 38–45, 2012.

8. Z. M. Fadlullah, N. Kato, R. Lu, X. Shen, and Y. Nozaki, "Toward secure targeted broadcast in smart grid," *IEEE Communications Magazine*, vol. 50, no. 5, pp. 150–156, 2012.

9. M. M. Fouda, Z. M. Fadlullah, N. Kato, R. Lu, and X. Shen, "A lightweight message authentication scheme for smart grid communications," *IEEE Transactions on Smart Grid*, vol. 2, no. 4, pp. 675–685, 2011.

10. R. Lu, X. Liang, X. Li, X. Lin, and X. Shen, "Eppa: An efficient and privacy-preserving aggregation scheme for secure smart grid communications," *IEEE Transactions on Parallel and Distributed Systems*, vol. 23, no. 9, pp. 1621–1631, 2012.

11. H. Li, X. Liang, R. Lu, X. Lin, and X. Shen, "Edr: an efficient demand response scheme for achieving forward secrecy in smart grid," in *Proc. GLOBECOM*, pp. 929–934, IEEE, 2012.

12. Y.-F. Chang and C.-C. Chang, "Enhanced anonymous auction protocols with freewheeling bids," in *Proc. AINA*, vol. 1, pp. 6–11, IEEE, 2006.

13. M.-J. Li, J. S.-T. Juan, and J. H.-C. Tsai, "Practical electronic auction scheme with strong anonymity and bidding privacy," *Information Sciences*, vol. 181, no. 12, pp. 2576–2586, 2011.

14. J. Shao, Z. Cao, X. Liang, and H. Lin, "Proxy re-encryption with keyword search," *Information Sciences*, vol. 180, no. 13, pp. 2576–2587, 2010.

15. B. Libert and J.-J. Quisquater, "The exact security of an identity based signature and its applications.," *IACR Cryptology ePrint Archive*, vol. 2004, p. 102, 2004.

16. H.-T. Liaw, W.-S. Juang, and C.-K. Lin, "An electronic online bidding auction protocol with both security and efficiency," *Applied mathematics and computation*, vol. 174, no. 2, pp. 1487–1497, 2006.

17. M. Wen, R. Lu, J. Lei, H. Li, X. Liang, and X. S. Shen, "Sesa: an efficient searchable encryption scheme for auction in emerging smart grid marketing," *Security and Communication Networks*, vol. 7, no. 1, p. 234–244, 2013.

18. Q. Liu, G. Wang, and J. Wu, "An efficient privacy preserving keyword search scheme in cloud computing," in *Proc. CSE*, vol. 2, pp. 715–720, IEEE, 2009.

19. M. Scott, "Efficient implementation of cryptographic pairings," in *[Online]. http://www.pairing-conference.org/2007/invited/Scottslide.pdf*, 2007.

20. S. D. Galbraith, K. G. Paterson, and N. P. Smart, "Pairings for cryptographers," *Discrete Applied Mathematics*, vol. 156, no. 16, pp. 3113–3121, 2008.

21. D.-J. Kang, B. H. Kim, and D. Hur, "Supplier bidding strategy based on non-cooperative game theory concepts in single auction power pools," *Electric power systems research*, vol. 77, no. 5, pp. 630–636, 2007.

22. D. Boneh, G. Di Crescenzo, R. Ostrovsky, and G. Persiano, "Public key encryption with keyword search," in *Proc. Eurocrypt*, pp. 506–522, Springer, 2004.

23. X. Lin, R. Lu, K. Foxton, and X. S. Shen, "An efficient searchable encryption scheme and its application in network forensics," in *Proc. E-Forensics*, pp. 66–78, Springer, 2011.

24. B. Zhang and F. Zhang, "An efficient public key encryption with conjunctive-subset keywords search," *Journal of Network and Computer Applications*, vol. 34, no. 1, pp. 262–267, 2011.

Chapter 3
Conjunctive Query over Encrypted Multidimensional Data

With the deployment of smart meters at individual households, smart grid can collect metering data of users' power consumption. However, users' power usage patterns would also be revealed. To preserve the users' privacy, metering data is mostly encrypted by cryptographic algorithms. When data mining is needed to support decision making or ensure reliability, to find useful information from the encrypted data is very important for smart grid. Most of the traditional keyword searching schemes rarely consider both users' data privacy and requesters' query privacy. In particular, the power system data in smart grid has multidimensional attributes; thus, how to query over the encrypted multidimensional data on all dimensions is a challenging issue in smart grid. To achieve finer grained conjunctive query, this chapter introduces an Efficient Conjunctive Query (ECQ) scheme. Specifically, the ECQ incorporates the idea of public key encryption and conjunctive keywords search to achieve conjunctive query without data and query privacy leakage. Security analysis demonstrates that the ECQ can achieve the security requirements, namely, data confidentiality, integrity and privacy, as well as query privacy.

3.1 Introduction

In smart grid, a short uploading interval is always desired to accurately reflect power usages. However, given the fact that the metering data of individual homes/factories is accumulated every 15 min, it is possible to infer the pattern of electricity consumption by individual users [1]. For example, the actions of the residents can be easily tracked by analyzing the smart meter data (gas, water, and electricity consumption). It is even possible to determine the presence/absence of residents or the number of people living in a household. In order to protect users' privacy, the sensitive data should be stored in an encrypted form. Lu et al. [1] design an efficient and privacy-preserving aggregation scheme, named EPPA, by using the homomorphic paillier cryptosystem technique. Li et al. [2] propose an efficient merkle tree based authentication scheme for smart grid. However, these privacy preserving schemes cannot support queries on encrypted data.

Public key encryption with keyword search (PEKS) [3] is a widely studied approach to achieve querying on encrypted data. Nevertheless, most of the existing schemes (such as [4, 5]) about PEKS focus only on the keyword search technique, with little attention to both data and query privacy protection in one scheme. Baek et al. [6] argue that PEKS and data encryption schemes need to be treated as a single scheme to securely provide PEKS service. Qin et al. [7] propose an efficient encryption scheme with one-dimension keyword search (EPPKS) for cloud computing by combining the ideas of partial decipherment with the PEKS. However, it is not quite secure because if the server is untrusted, the partial decipherment will leak partial information of users' data. Furthermore, the EPPKS cannot support conjunctive keyword searches on multiple dimensions. The reason is that power system data usually has multi-dimensional attributes. Taking into account that all these dimensions allow finer grained query, when a conjunctive query on multiple dimensions is posted, the EPPKS will have to process the query on every dimension separately.

To protect data privacy and save communication and computation overhead, in this chapter, we address the conjunctive query supporting subset keywords problem in smart grid. This scheme supports conjunctive query on encrypted multi-dimensional data by giving conjunctive keywords on multiple dimensions. The main contents of this chapter are two-fold.

1. Firstly, An efficient conjunctive query (ECQ) scheme will be constructed, which considers both data and query privacy preservation when querying over the encrypted multi-dimensional data. Requesters can get matched results by giving conjunctive keywords on multiple dimensions. Compared with the EPPKS [7], the ECQ is more efficient in terms of user's computation cost and total communication cost.
2. Secondly, we analyze the security strength and privacy preservation ability of the ECQ. The analysis results show that the ECQ can protect the data confidentiality and integrity, as well as data and query privacy. Even if the server is compromised, our proposed ECQ scheme can still protect data privacy without any information leakage, which is more secure than the EPPKS [7].

The remainder of this paper is structured as follows. In Sect. 3.2, we introduce our system model, security requirements and design goals. The ECQ scheme is described in Sect. 3.3; followed by its security analysis and performance evaluation in Sect. 3.4. In Sect. 3.5, we review the related works. Finally, we draw our summary in Sect. 3.6.

3.2 System Model, Security Requirements and Design Goal

In this section, we formalize the system model, identify the security requirements and our design goals.

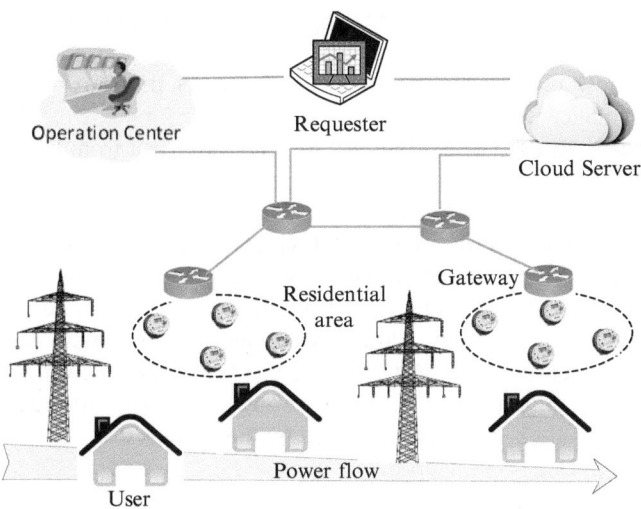

Fig. 3.1 System model of ECQ

3.2.1 System Model

In the system model, we mainly focus on how to query over encrypted multi-dimensional data for smart grid by using conjunctive keywords. Specifically, we consider a typical residential area, as shown in Fig. 3.1, which comprises an operation center, a cloud server (CS), several local gateways (GA), a requester, and a large number of smart meters (SM) related to the corresponding residential users $\mathbf{U} = \{U_1, U_2, \ldots, U_v\}$.

For data outsourcing, the power usage data of residential users will be stored on the cloud server forwarded by the GA. This is a promising approach to relieve the operation center from the burden of large amounts of data storage and maintenance, and executing computations and queries using the servers' computational capabilities [8]. Requesters, such as market analysts and utility companies, will be able to query the cloud server to find useful information. The operation center can be considered as a trusted authority (TA), and can bootstrap the whole system. Specifically, the TA generates and distributes keys for residential users and authorized requesters.

3.2.2 Security Requirements

We define the security requirements for the efficient conjunctive query (ECQ) scheme, and will show the fulfillment of these requirements after presenting the design details.

- *Data Confidentiality and Integrity*: The power system data and queries should not be known and changed by malicious users or unauthorized users. That is, if an adversary $\mathscr{A}$ maliciously modifies data on the CS, the power demand forecast and other policies will be misled.
- *Data privacy*: As users usually prefer to keep their data from being exposed to others, including the CS, the most important concern is preserving data privacy. It means that only requesters with correct secret key can obtain the correct data when their query keywords are satisfied with those in the encrypted data.
- *Query privacy*: As requesters usually prefer to keep their queries from being exposed to others, thus, the biggest concern is to hide their queries into trapdoors to protect the query privacy. Otherwise, if the query includes some sensitive information, such as *"city hall"*, then the CS could know the requester is querying some important locations' metering data. Then, the requester or the query results could be traced or analyzed by the curious CS.

3.2.3 Design Goal

Under the aforementioned system model and security requirements, our design goal is to develop an efficient conjunctive query scheme for smart grid, achieving data security and efficient performance as follows.

- The proposed ECQ scheme should remain secure to meet the security requirements: the data confidentiality, integrity, and data and query privacy. Otherwise, the user-specific data leakage can lead to criminal targeting of homes.
- The proposed scheme should achieve conjunctive query efficiency over encrypted multidimensional data in terms of communication and computation overhead.

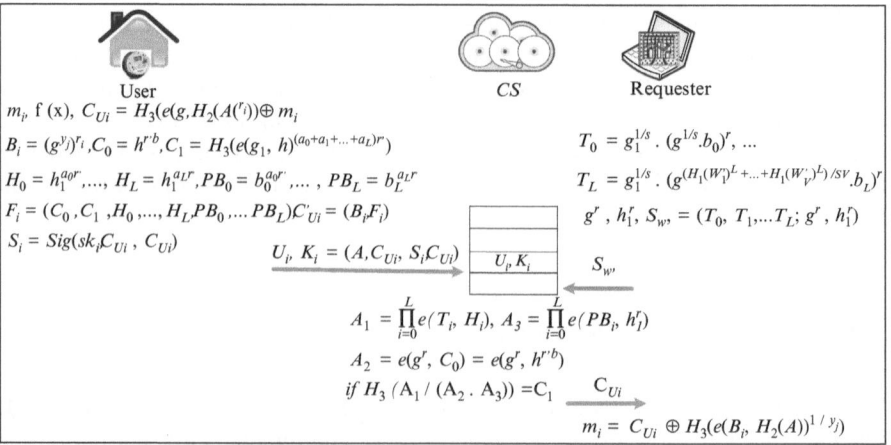

Fig. 3.2 Dataflow of ECQ

3.3 The ECQ Scheme

In this section, we show the construction of the efficient conjunctive query scheme on encrypted data for smart grid. It mainly consists of the following four phases: registration phase, data and tags encryption phase, conjunctive query phase, and data recovery phase. The ECQ incorporates and modifies the idea of keywords search approach [5] and the proxy re-encryption with keyword search [9]. The dataflow is shown in Fig. 3.2.

In system initialization phase, the TA can bootstrap the whole system and assign the key materials. TA first generates (q, g, G_1, G_2, e) by running $\mathscr{G}en(1^k)$, where G_1 and G_2 are two cyclic multiplicative groups of prime order q; and g is a generator of group G_1. An identity based signature algorithm $Sig(\cdot)$ [10] is needed in our scheme. Furthermore, we need three hash functions $H_1 : \{0, 1\}^* \to G_1$, $H_2 : \{0, 1\}^* \to G_1$ and $H_3 : G_2 \to \{0, 1\}^n$ for some n. TA publishes the system parameters as $(q, g, G_1, G_2, e, H_1, H_2, H_3)$.

3.3.1 Registration Phase

In our scheme, there are n users $\mathbf{U} = \{U_1, U_2, \ldots, U_n\}$ and m requesters $\mathbf{SE} = \{SE_1, SE_2, \ldots, SE_m\}$ in the system. The TA randomly chooses a master key $s \in Z_q^*$, and computes the corresponding public key $pk = g^s$. Thus, (g^s, s) is the public/private key pair of the TA. When each SE_j registers, the TA picks a random number $y_i \in Z_q^*$ and sets $pb_j = g^{y_j}$. (pb_j, y_j) is SE_j's public/private key pair. For each U_i, the TA assigns an ID-based key pair $(H_1(ID_{U_i}), H_1^s(ID_{U_i}))$ to it, denoted as (vk_i, sk_i). Then, TA also chooses parameters $g_1, b_0, b_1, \ldots, b_L \in G_1$, $h \in G_2$ and computes $h_1 = h^s$.

3.3.2 Data and Tags Encryption Phase

Without loss of generality, we assume that there are L dimensions in the power system data, and each dimension will be characterized by a keyword (e.g., user type, date etc.). Thus, U_i selects a keyword set $W = \{w_1, w_2, \ldots, w_L\}$ to characterize its data and encrypts them as following steps.

1. U_i denotes its identity-based public key as $A = vk_i$, and keeps the private key sk_i secretly.
2. Let a L-dimensional power data be generated as $m_i = (pout_i, pin_i, pa_i, Ts_i, \triangle)$, where pin_i denotes the energy amount bought by U_i from the grid; $pout_i$ is energy amount sold by U_i to the grid; pa_i is U_i's total payment; Ts_i is the timestamp of the data and $\triangle$ is U_i's other information. Then, U_i selects a random $r_i \in Z_q^*$ and computes $C_{Ui} = H_3(e(g, H_2(A)^{r_i})) \oplus m_i$.
3. U_i computes $B_i = (g^{y_j})^{r_i}$ and selects the elements $a, b, r' \in Z_q$. Then, it constructs a L-degree polynomial:

$$f(x)$$

$$= a.(x - H_1(w_1))(x - H_1(w_2)) \ldots (x - H_1(w_L)) + b \qquad (3.1)$$

$$= a_L x^L + \ldots + a_1 x + a_0.$$

In this way, $H_1(w_1), \ldots, H_1(w_L)$ are L roots of the equation $f(x) - b = 0$. With the parameters $(a_0, a_1, \ldots, a_L, b)$, U_i then computes a tag F_i:

$$
\begin{aligned}
C_0 &= h^{r'b}, \\
C_1 &= H_3(e(g_1, h)^{(a_0 + a_1 + \ldots + a_L)r'}), \\
H_0 &= h_1^{a_0 r'}, \cdots H_L = h_1^{a_L r'}; \\
PB_0 &= b_0^{a_0 r'}, \cdots PB_L = b_L^{a_L r'}.
\end{aligned}
\qquad (3.2)
$$

We denote $F_i = (C_0, C_1, H_0, \ldots, H_L, PB_0, \ldots PB_L)$, $C'_{Ui} = (B_i, F_i)$.

4. U_i generates a signature $S_i = Sig(sk_i, C_{Ui}, C'_{Ui})$ by using its secret key sk_i.
5. U_i sends the encrypted data $K_i = (A, C_{Ui}, S_i, C'_{Ui})$ to the GA. The GA forwards it to the CS.
6. The CS will store this information received from U_i as a record (U_i, K_i) in its database.

3.3.3 Conjunctive Query Phase

TA computes two parameters $\alpha = g^{1/s}$, $\beta = g_1^{1/s}$ and secretly sends them to SE_j. If SE_j needs to query the data over v dimensions. It will generate query Q_j with keywords set $Q_j = \{w'_1, w'_2, \ldots, w'_v\}$. Then SE_j generates a trapdoor $S_{w'}$ and sends it to the CS. Without loss of generality, we assume, $\{w'_1, w'_2, \ldots, w'_v\}$ can be any subset of $\{w_1, w_2, \ldots, w_L\}$.

1. Trapdoor generation:
   SE_j picks a number $r \in Z_q$ and generates a trapdoor $S_{w'}$ on the query Q_j as:

$$T_0 = g_1^{1/s}.(g^{1/s}.b_0)^r,$$

$$\ldots$$

$$T_i = g_1^{1/s}.(g^{(H_1(w'_1)^i + \ldots + H_1(w'_v)^i)/sv}.b_i)^r,$$

$$\ldots$$

$$T_L = g_1^{1/s}.(g^{(H_1(w'_1)^L + \ldots + H_1(w'_v)^L)/sv}.b_L)^r,$$

$$g^r, h_1^r. \qquad (3.3)$$

Let $S_{w'} = (T_0, T_1, \ldots T_L; g^r, h_1^r)$. Then, SE_j sends $S_{w'}$ to the CS.

2. Query on the encrypted data:

For each encrypted data K_i in CS's database, the CS tests if C'_{Ui} satisfies the SE_j's requirement $S_{w'}$:

(1) Data verification:

- The CS verifies signature S_i on data (C_{Ui}, C'_{Ui}) with respect to U_i's public key A.
- If the signature verification fails, the CS checks the next data in its database; else the CS goes on testing.

(2) Now, the CS computes some parameters as follows:

$$A_1 = \prod_{i=0}^{L} e(T_i, H_i),$$

$$A_2 = e(g^r, C_0) = e(g^r, h^{r'b}),$$

$$A_3 = \prod_{i=0}^{L} e(PB_i, h_1^r). \qquad (3.4)$$

(3) The CS tests if $H_3(A_1/(A_2.A_3)) = C_1$. If so, C_{Ui} will be stored in a data array X[]; if not, C_{Ui} will be rejected.

3. The CS sends results in data array X[] to SE_j.

3.3.4 Data Recovery Phase

Upon receiving data array X[] from the CS, the SE_j can decrypt each C_{Ui} in X[] as $m_i = C_{Ui} \oplus H_3(e(B_i, H_2(A))^{1/y_j})$ by using its secret key y_j, otherwise, C_{Ui} will be discarded. The decryption is as follows.

$$C_{Ui} \oplus H_3(e(B_i, H_2(A))^{1/y_j})$$
$$= H_3(e(g, H_2(A)^{r_i})) \oplus m_i \oplus H_3(e((g^{y_j})^{r_i}, H_2(A))^{1/y_j})$$
$$= H_3(e(g, H_2(A)^{r_i})) \oplus m_i \oplus H_3(e(g, H_2(A)^{r_i}))$$
$$= m_i$$

3.4 Performance Analysis

3.4.1 Security Analysis

Since the idea of the ECQ is based on schemes [5,9], the provable security of the ECQ can be easily reduced to them. Thus, the provable security analysis of the ECQ can reference to schemes [5,9]. In this subsection, we just analyze the security properties of the proposed ECQ scheme.

- *The confidentiality and integrity of the users' data is achieved in the proposed ECQ scheme*: In ECQ, each data is encrypted by its secret number r_i as $C_{Ui} = H_3(e(g, H_2(A)^{r_i})) \oplus m_i$. Anyone who does not know r_i, can not recover m_i from the ciphertext C_{Ui}. Furthermore, the data is signed by identity based signature scheme [10]. Since the signature $S_i = Sig(sk_i, C_{Ui}, C'_{Ui})$ is provably secure, the source authentication and data integrity can be guaranteed.
- *The users' data privacy and the requesters' query privacy are also achieved in the proposed ECQ scheme*: On one hand, as shown in Eqs. (3.1) and (3.3), the keywords hash values are used to calculated the data tag and trapdoors and they are in the exponent of g. Anyone can not recover original keywords w_i with the tag C'_{Ui} because of the one-way property of hash function.

  On the other hand, when the CS tests the query on the encrypted data, it also cannot know anything about the keywords in the query, except for whether the test is successful or not. The correctness of the test $H_3(A_1/(A_2.A_3)) = C_1$ is as follows. Since

$$
\begin{aligned}
A_1 &= \prod_{i=0}^{L} e(g_1^{1/s} \cdot (g^{(H_1(w'_1)^i + \dots + H_1(w'_v)^i)/sv} \cdot b_i)^r, h^{a_i.s.r'}) \\
&= e(g_1, h)^{(a_0 + a_1 + \dots + a_L).r'} \\
&\quad \cdot e(g^r, h^{r'})^{1/v(\sum_{i=0}^{L} a_i(H_1(w'_1)^i + \dots + H_1(w'_v)^i))} \\
&\quad \cdot \prod_{i=0}^{L} e(b_i^{a_i.s.r'}, h^r),
\end{aligned}
\tag{3.5}
$$

if $H_1(w'_1), \dots, H_1(w'_v)$ are the roots of equation $f(x) - b = 0$, which means that all the keywords in the trapdoor are the subset of the keywords in the tag, i.e. $\{w'_1, w'_2, \dots, w'_v\} \subseteq W$, then the CS can get the equation:

$$
\begin{aligned}
&1/v(\sum_{i=0}^{L} a_i(H_1(w'_1)^i + \dots + H_1(w'_v)^i)) \\
&= 1/v(\sum_{i=0}^{L} a_i H_1(w'_1)^i + \dots + \sum_{i=0}^{L} a_i H_1(w'_v)^i) \\
&= b.
\end{aligned}
\tag{3.6}
$$

Therefore,

$$A_1 = e(g_1, h)^{(a_0 + a_1 + \ldots + a_L)r'}$$

$$\cdot e(g^r, h^{r'b}) \cdot \prod_{i=0}^{L} e(b_i^{a_i.s.r'}, h^r)$$

$$= e(g_1, h)^{(a_0 + a_1 + \ldots + a_L)r'} \cdot A_2 \cdot A_3. \tag{3.7}$$

Notice that the keywords in the trapdoor can be listed in any order when the SE_j computes the trapdoor. Substitute Eq. (3.7) into $H_3(A_1/(A_2.A_3))$, we have C_1. Thus, the test is successful, and the CS cannot know anything about the keywords in the query.

- *The proposed ECQ can remain secure even when the server is compromised*: If the CS is malicious, the user's data privacy will be leaked in EPPKS [7], because the partial decipherment key computed by the server will leak the partial information of the encrypted data. Whereas, in the proposed ECQ scheme the confidentiality of the encrypted data is always preserved even if the cloud server is malicious. From the above analysis, we can see that the ECQ can remain secure even when the server is compromised.

3.4.2 Performance Evaluation

In this subsection, we will evaluate the ECQ in terms of the computation and communication complexities.

Computation: In the ECQ, the computation tasks include pairing operations and exponentiation operations etc. For simplicity of description, the pairing and exponentiation operations are denoted as C_p and C_e, respectively. For instance, there are L dimensions in a power usage data m, and m will be characterized by L keywords in a keyword set W. U_i will encrypt m into a ciphertext C_{Ui}, and compute a tag C'_{Ui} on W. Therefore, it costs 4 pairings and $(2L + 4)$ exponentiation operations for U_i to generate a data encryption $(A, C_{Ui}, S_i, C'_{Ui})$, i.e., $4C_p + (2L + 4)C_e$. Then, SE_j needs $(2L + 2)$ exponentiation operations to compute a trapdoor $S_{w'}$. After receiving the trapdoor $S_{w'}$ from SE_j, the CS needs 2 pairings to verify the signature, $(2L + 3)$ pairings and 3 exponentiation operations to test SE_j's query. If there is a satisfactory result, SE_j requires 1 pairing operation and 1 exponentiation operation to decrypt the ciphertext.

In comparison, for a L-dimensional data, in the traditional approach (e.g. EPPKS [7]) a user should generate $(L + 2)$ pairings and $(2L + 4)$ exponentiation operations to generate a data encryption with L tags on L dimensions. That totally costs $(L + 2)C_p + (2L + 4)C_e$. Since EPPKS [7] only can do 1 keyword query at a time, for a query with L keywords, the requester needs to compute L trapdoors, which needs L exponentiation operations. And the sever needs to test L times. Every

Table 3.1 Comparison of computation complexity

	ECQ [11]	EPPKS [7]
User	$4C_p + (2L+4)C_e$	$(L+2)C_p + (2L+4)C_e$
CS	$(2L+3)C_p + 3C_e$	$(2L+2)C_p + 2C_e$
Requester	$C_p + (2L+3)C_e$	$(L+1)C_e$

time, the server needs 2 pairings to test whether a given tag matches a trapdoor or not. Thus, the sever totaly needs $2LC_p$ to test all of the trapdoors. If there is a satisfactory result, the sever needs $2C_p + 2C_e$ more computation overhead to get a partial decipherment. At last, it will take the requester C_e to recover the real data.

Detailed experiments are also conducted on a Pentium IV 3-GHz system to study the execution time [12]. For G_1 over the FST curve, a single C_e in G_1 with 161 bits costs 1.1 ms, and the C_p costs 3.1 ms. From Table 3.1, Figs. 3.3 and 3.4 we can see that a user in the ECQ needs much less computation overhead than in the EPPKS [7]. Although the CS and requester need a little more computation overhead in the ECQ scheme than that in EPPKS [7], the ECQ needs less total computation overhead when the number of dimensions larger than 7. Thus, the ECQ is more suitable to the smart grid where there are more users and less requesters.

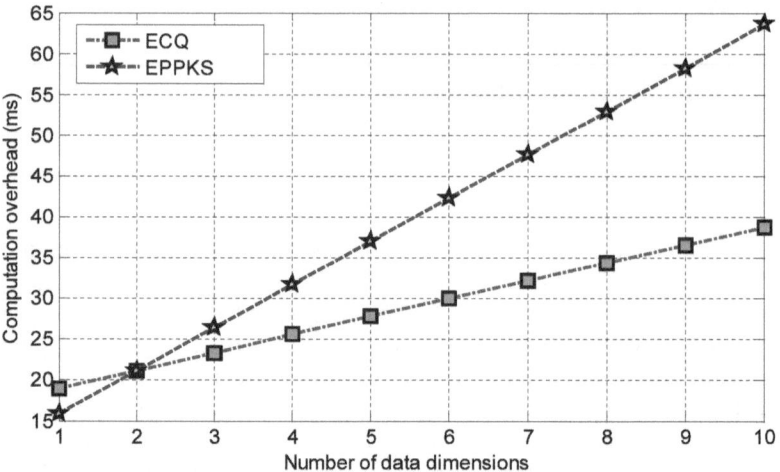

Fig. 3.3 Computation overhead of each user

Communication: Most pairing-based cryptosystems need to work in a subgroup of the elliptic curve $E(F_q)$. If we represent elliptic curve points by using the point compression approach [13], the length of the elements in G_1 and G_2 will be roughly 161-bit and 1,024-bit, respectively. We assume that SHA-1 is used to compute the hash function, which yields a 160-bit output. The communications among the three entities of the proposed ECQ can be divided into three parts, user-to-CS, requester-to-CS and CS-to-requester communications.

We first consider the user-to-CS communication in the ECQ. In the data encryption phrase, users generate their encrypted data and tags, then they deliver them to the CS. The data report is in the form of $K_i = (A, C_{Ui}, S_i, C'_{Ui})$, where, A is a G_1 element; C_{Ui} is a hash value; signature S_i includes 2 G_1 elements [10]; C'_{Ui} includes $(2L + 1)$ G_1 elements and 1 hash value; thus, the size of K_i should be $161 * (2L + 1) + 160 * 2$ bits. In the requester-to-CS communication, each requester deliveries a trapdoor $S'_w = (T_0, T_1, \ldots T_L, g^r, h^r_1)$ to the CS, which includes $(L + 2)$ G_1 elements, i.e., $161(L + 2)$ bits. In the CS-to-requester communication, the CS will reply a 160-bit ciphertext C_{Ui} to SE_j.

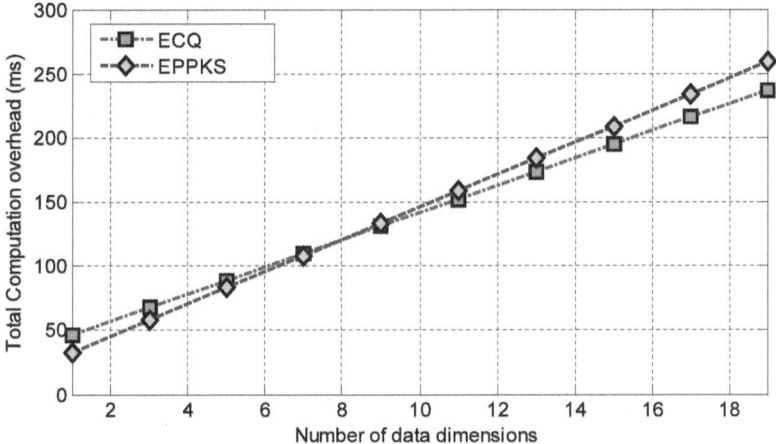

Fig. 3.4 Total computation overhead

Table 3.2 Comparison of communication complexity(bits)	ECQ	EPPKS [7]
User-to-CS	$161 * (2L + 1) + 320$	$481 + 160L$
Requester-to-CS	$161(L + 2)$	$160L$
CS-to-requester	160	$160L + 1,405$

In the EPPKS [7], the user sends a message $(C_m, C_{w_i}, i = 1, \ldots L)$ to the server, which includes a G_1 element and $L + 2$ hash elements. Its size is $(481 + 160L)$ bits. Then, a 160-bit long trapdoor will be sent from the requester to the server. L trapdoors are $160L$ bits. In the server-to-requester communication, the server replies $(C_m, C_\rho, C_{w_i}, i = 1, \ldots L)$ to the requester. Here, C_ρ is a G_2 element, C_m includes a G_1 element and 2 hash elements, C_{w_i} is a hash value. The size of the reply is $(160L + 1,405)$ bits.

Table 3.2, Figs. 3.5 and 3.6 show the comparison of communication overhead between the ECQ and EPPKS [7]. The ECQ needs much less communication overhead than the EPPKS [7], especially during CS-to-requester communication and total communication overhead.

3.5   Related Works

Boneh et al. [3] firstly introduced the concept of public key encryption with keyword search (PEKS). Later, many papers were published to extend it. In [14], Byun et al. pointed out a drawback about the scheme (keywords guessing attack), Rhee et al. [15] and Camenisch et al. [16] gave out the way to cope with this attack accordingly. The notion of public key encryption with conjunctive field keyword search was proposed by Park et al. [17]. The "field" means that to confirm the number of keywords in the message and the trapdoor are similar with two assumptions.

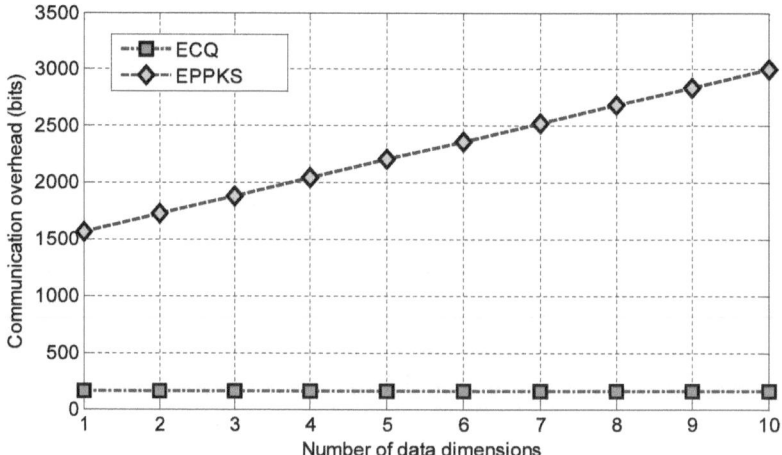

Fig. 3.5 Communication overhead of CS-to-requester

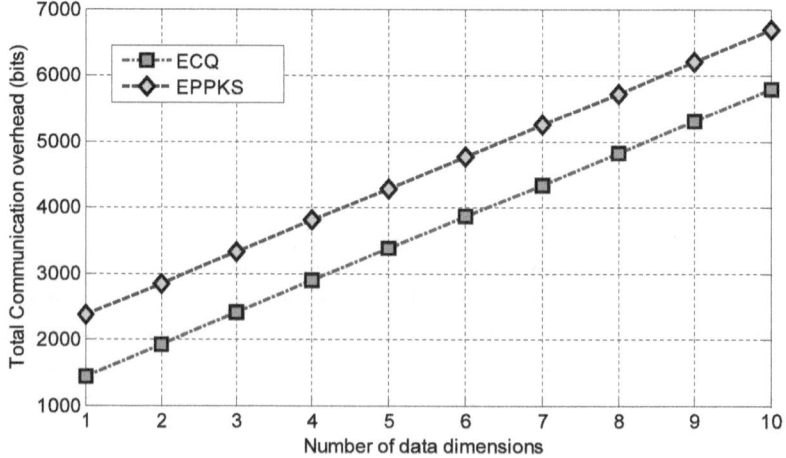

Fig. 3.6 Total communication overhead

For example, in an email system, the keyword field could be confirmed as: "From", "To" and "When". Either trapdoor or the message has at least three keywords, also must in this order. This restriction makes the conjunctive keyword search unpractical. Byun et al. [18] and Hwang et al. [19] improved the efficiency of the conjunctive keyword search, while both of them still do not support subset keyword search.

In 2007, Boneh and Waters [20] constructed public-key systems that support comparison queries on encrypted data as well as more general queries such as subset queries. At first, they constructed a Hidden Vector Encryption scheme. Then based on this scheme, subset queries could be realized. However, in the keyword search environment, there exist some flaws. The efficiency is low, and the keywords order should be fixed, meanwhile they mapped the key words to a index set. Zhang et al. proposed a scheme PECSK [5] to support conjunctive-subset keywords search. But it is only one keyword search scheme. The data encryption should use another encryption algorithm to do so. As far as we know, in cloud computing, Liu et al. [7] presented a privacy preserving keyword search scheme, named EPPKS. It is one of the few schemes which integrates both the message encryption and keyword search properties. However, when the server finds a tag matching the trapdoor in EPPKS, the server has to compute an intermediate result to help the user to recover the message, which not only can cost communication and computation overhead but also cause the security vulnerability.

3.6 Summary

In this chapter, we have developed an efficient conjunctive query scheme over encrypted multi-dimensional data in smart grid. In particular, it achieves both data and query privacy-preserving. Requesters, such as utility companies or marketing managers, can retrieve useful information by querying with conjunctive keywords. Security analysis demonstrates that our ECQ can achieve data confidentiality and integrity, data and query privacy. Simulation results show that our scheme can reduce the users' computation cost and total communication cost. However, range query is also very important for smart grid in some applications. For example, in the smart grid auction market, energy providers may use the range of price keyword to filter out the bids with reasonable price; utility managers may use the range of dates or locations to query power consumption data for decision making etc. Hence, in the next chapter, we intend to study range queries over encrypted multidimensional data in smart grid.

References

1. R. Lu, X. Liang, X. Li, X. Lin, and X. Shen, "Eppa: An efficient and privacy-preserving aggregation scheme for secure smart grid communications," *IEEE Transactions on Parallel and Distributed Systems*, vol. 23, no. 9, pp. 1621–1631, 2012.
2. H. Li, R. Lu, L. Zhou, B. Yang, and X. Shen, "An efficient merkle-tree-based authentication scheme for smart grid," *IEEE Systems Journal*, vol. pp, no. 99, pp. 1–9, 2013.
3. D. Boneh, G. Di Crescenzo, R. Ostrovsky, and G. Persiano, "Public key encryption with keyword search," in *Proc. Eurocrypt*, pp. 506–522, Springer, 2004.
4. M. Wen, R. Lu, K. Zhang, J. Lei, X. Liang, and X. Shen, "Parq: A privacy-preserving range query scheme over encrypted metering data for smart grid," *IEEE Transactions on Emerging Topics in Computing*, vol. 1, no. 1, pp. 178–191, 2013.
5. B. Zhang and F. Zhang, "An efficient public key encryption with conjunctive-subset keywords search," *Journal of Network and Computer Applications*, vol. 34, no. 1, pp. 262–267, 2011.
6. J. Baek, R. Safavi-Naini, and W. Susilo, "On the integration of public key data encryption and public key encryption with keyword search," in *Proc. ISC*, pp. 217–232, Springer, 2006.
7. Q. Liu, G. Wang, and J. Wu, "An efficient privacy preserving keyword search scheme in cloud computing," in *Proc. CSE*, vol. 2, pp. 715–720, IEEE, 2009.
8. B. Hore, S. Mehrotra, M. Canim, and M. Kantarcioglu, "Secure multidimensional range queries over outsourced data," *The International Journal on Very Large Data Bases*, vol. 21, no. 3, pp. 333–358, 2012.
9. J. Shao, Z. Cao, X. Liang, and H. Lin, "Proxy re-encryption with keyword search," *Information Sciences*, vol. 180, no. 13, pp. 2576–2587, 2010.
10. B. Libert and J.-J. Quisquater, "The exact security of an identity based signature and its applications.," *IACR Cryptology ePrint Archive*, vol. 2004, p. 102, 2004.
11. M. Wen, R. Lu, X. Liang, H. Li, and X. S. Shen, "Ecq: An efficient conjunctive query scheme over encrypted multidimensional data in smart grid," in *Proc. Globcom*, to appear.
12. M. Scott, "Efficient implementation of cryptographic pairings," in *[Online]. http://www.pairing-conference.org/2007/invited/Scottslide.pdf*, 2007.
13. S. Galbraith, "Pairings," *LondonN Mathematical Society Lecture Note Series*, vol. 317, p. 183, 2006.
14. J. W. Byun, H. S. Rhee, H.-A. Park, and D. H. Lee, "Off-line keyword guessing attacks on recent keyword search schemes over encrypted data," in *Proc. SDM*, pp. 75–83, Springer, 2006.
15. H. S. Rhee, J. H. Park, W. Susilo, and D. H. Lee, "Trapdoor security in a searchable public-key encryption scheme with a designated tester," *Journal of Systems and Software*, vol. 83, no. 5, pp. 763–771, 2010.
16. J. Camenisch, M. Kohlweiss, A. Rial, and C. Sheedy, "Blind and anonymous identity-based encryption and authorised private searches on public key encrypted data," in *Proc. PKC*, pp. 196–214, Springer, 2009.
17. P. Golle, J. Staddon, and B. Waters, "Secure conjunctive keyword search over encrypted data," in *Proc. ACNS*, pp. 31–45, Springer, 2004.
18. J. W. Byun, D. H. Lee, and J. Lim, "Efficient conjunctive keyword search on encrypted data storage system," in *Proc. EuroPKI*, pp. 184–196, Springer, 2006.
19. R. Curtmola, J. Garay, S. Kamara, and R. Ostrovsky, "Searchable symmetric encryption: improved definitions and efficient constructions," in *Proc. CCS*, pp. 79–88, ACM, 2006.
20. D. Boneh and B. Waters, "Conjunctive, subset, and range queries on encrypted data," in *Proc. TCC*, pp. 535–554, 2007.

Chapter 4
Range Query over Encrypted Metering Data for Financial Audit

Smart grid, envisioned as an indispensable power infrastructure, is featured by real-time and two-way communications. However, how to securely retrieve and audit the communicated metering data for validation testing is still challenging for smart grid. In this chapter, we introduce a novel privacy-preserving range query scheme over encrypted metering data, named PaRQ, to address the privacy issues in financial auditing for smart grid. The PaRQ allows a residential user to store metering data on a cloud server in an encrypted form. When financial auditing is needed, an authorized requester can send its range query tokens to the cloud server to retrieve the metering data. Specifically, the PaRQ constructs a hidden vector encryption (HVE) based range query predicate to encrypt the searchable attributes and session keys of the encrypted data. Meanwhile, the requester's rang query can be transferred into two query tokens, which are used to find the matched query results.

4.1 Introduction

Smart grid has emerged as a new concept and a promising solution for intelligent electricity generation, transmission, distribution and control. The use of robust two-way communications and distributed computing technology improves the efficiency and reliability of power delivery and usage [1]. Currently, many utility companies begin to use smart grid information systems to collect real-time metering data at their control centers, via a reliable communication network deployed in parallel to the power transmission and distribution grid [2], as shown in Fig. 4.1. In the smart grid information system, smart meters are deployed at residential users' premises as two-way communication devices, which periodically record the power consumption and report their metering data to a local area gateway, e.g., a wireless access point (AP). The gateway then collects and forwards data to a control center. Additionally, metering data in smart grid information systems should be periodically audited to ensure that the billing and pricing statements are presented fairly [3]. Specifically, requesters, such as market analysts, are endowed with the task of

M. Wen et al., *Querying over Encrypted Data in Smart Grids*, SpringerBriefs in Computer Science, DOI 10.1007/978-3-319-06355-3_4, © The Author(s) 2014

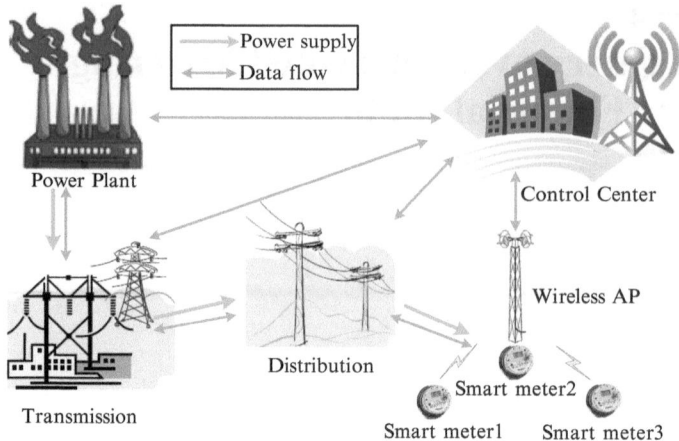

Fig. 4.1 The conceptional smart grid architecture

querying smart grid information systems for auditing, analysis, accounting or tax-related activities [4]. Thus, to prevent the private and sensitive information in the metering data from disclosure, data confidentiality and privacy should be achieved in financial audit for smart grid.

However, the metering data in smart grid are surging from 10,780 terabytes (TB) in 2010 to over 75,200 TB in 2015 [5], which is far beyond the control center's data management capability. Outsourcing data to cloud servers is a promising approach to relieve the control center from the burden of such a large amount of data storage and maintenance. In this approach, users can store their data on cloud servers and execute computation and queries using the servers' computational capabilities [6]. Nevertheless, cloud servers might be untrusted, and intentionally share sensitive data with the third parties for commercial purposes. Therefore, data confidentiality is important in financial audit for smart grid.

In addition, privacy concerns raise in financial auditing. For instance, utility usage patterns within short intervals may reveal the users' regular daily activities [7]. In particular, data from a single house would reveal the activities of the residents, e.g., when the individual resident is at home, when he/she is watching TV [2]. If an attacker can query these data, data privacy might be violated. Therefore, users' data confidentiality and privacy should be protected and only authorized requesters can query the metering data.

From the requester's perspective, the requester, who manages the data query for financial auditing, needs to frequently query the metering data by using date ranges and/or geographic regions etc. If the query is sensitive, the requesters may prefer to keep their queries from being exposed to servers. As a result, how to operate such range queries with guaranteed query privacy is also significant for smart grid.

In this chapter, a Privacy-preserving Range Query (PaRQ) scheme over encrypted metering data will be presented for smart grid. The PaRQ addresses the data confidentiality and privacy problem by introducing an HVE technique. The main contents of this paper are twofold.

- Firstly, a range query predicate based on the HVE is constructed. Specifically, the session keys and the searchable attributes of the encrypted data are hidden in the HVE based range query predicate. When a requester query the cloud server, the session keys, whose encryption vectors are satisfied with the range query vectors, are released to the requester, for decrypting the encrypted metering data.
- Secondly, we analyze the security strengths and evaluate the performance of the PaRQ. Security analysis demonstrates that the PaRQ can achieve user's data confidentiality and privacy, as well as requester's query privacy. Performance evaluation results show that our PaRQ can reduce the communication and computation overhead, and shorten the response time.

The remainder of this paper is organized as follows. In Sect. 4.2, we introduce our system model, security requirements and our design goal. Then, in Sect. 4.3, we present our PaRQ scheme, followed by its security analysis and performance evaluation in Sects. 4.4 and 4.5, respectively. In Sect. 4.6, we investigate the related works. Finally, we draw the summary of this chapter in Sect. 4.7.

4.2 System Model, Security Requirements and Design Goal

In this section, we formalize the system model, and identify the security requirements and our design goals.

4.2.1 System Model

Our focus is on how to outsource residential users' metering data to a cloud server in encrypted form and how to operate a range query over the encrypted metering data with the help of the control center (CC). Specifically, we consider a typical residential area, as shown in Fig. 4.2, which is composed of a CC, two cloud servers: the CS_1 and CS_2, a requester S and some residential users $\mathbf{U} = \{U_1, U_2, \ldots, U_v\}$.

A residential user is the data owner, who encrypts his data by using a secret session key before outsourcing the data to the CSs. There are two cloud servers: Cloud Server 1 (CS_1) stores data ciphertexts; Cloud Server 2 (CS_2) stores session key's ciphertexts and indexes. Both servers are semi-trusted, honest but curious. We assume that either the CS_1 or CS_2 might be compromised and controlled by an adversary seeking to link users' ciphertexts with their keys, but the adversary cannot control both CSs. The control center is a trusted proxy (it operates on behalf of the utility companies), which can help users to deposit their data to cloud servers and generate query tokens for requesters to retrieve data from the servers. The requester can query the encrypted data on the cloud servers by depositing his entitling tokens to the CS_2.

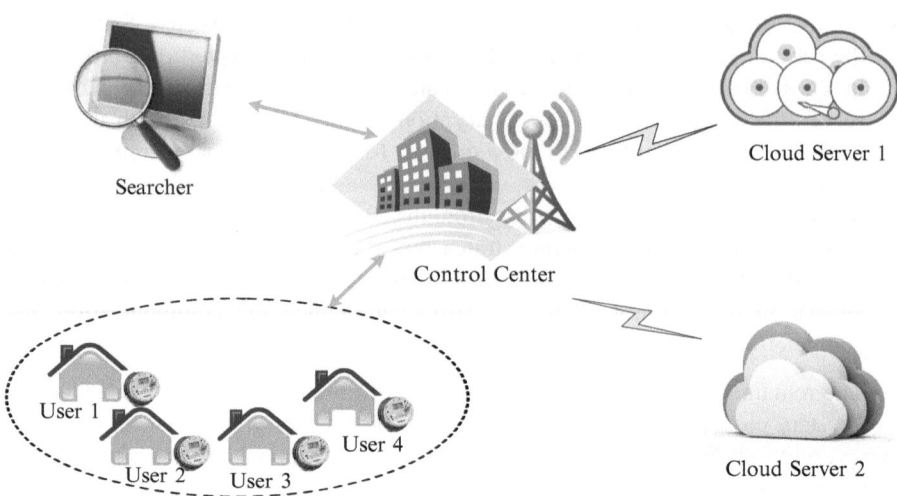

Fig. 4.2 System model of PaRQ

The CC consists of two main components: a ciphertext forwarder, and a query translator which always operates within the secure environment. The forwarder on the CC needs to add a unique index to the data ciphertexts and the session key's ciphertexts. To preserve the query privacy, the requester's query needs to be translated into two tokens, so that the CS_2 can evaluate this query without disclosing its real value.

4.2.2   Security Requirements

We identify the security requirements for our PaRQ. In our security model, the CC is trustable, and residential users $\mathbf{U} = \{U_1, U_2, \ldots, U_v\}$ are honest as well. However, there exists an adversary $\mathscr{A}$ in the system intending to eavesdrop and invade the database on cloud servers to steal the individual users' reports. In addition, $\mathscr{A}$ can also launch some active attacks to threaten the data privacy and query privacy. Therefore, in order to prevent $\mathscr{A}$ from learning the users' data and to detect its malicious actions, the following security requirements should be satisfied in range query applications for smart grid.

- *Data Confidentiality*: The residential user can utilize symmetric or asymmetric cryptography to encrypt the data before outsourcing, and successfully prevent the unauthorized entities, including eavesdroppers and cloud servers, from prying into the outsourced data.
- *Data privacy*: Individual residential users' data should not be accessed by unauthorized requesters. It means that only requesters with authorized query

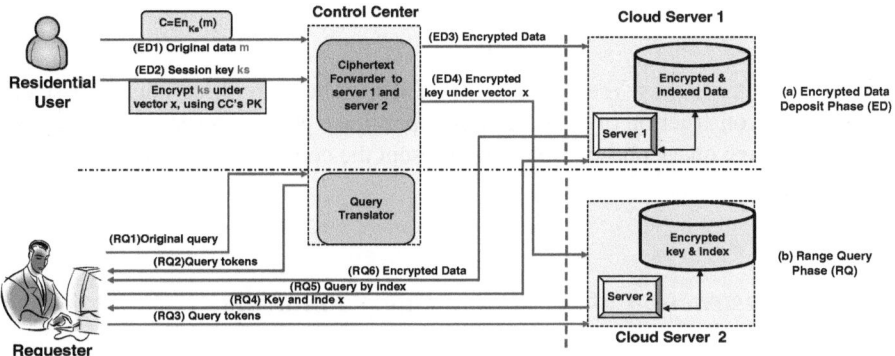

Fig. 4.3 Data query procedures

tokens can access the CS_2, and they can obtain the correct session keys when their query vectors in the tokens are satisfied with the encryption vectors. Thus, only the authorized requester can decrypt the encrypted metering data.

- *Query privacy*: As requesters usually prefer to keep their queries from being exposed to others, thus, the biggest concern is to hide their queries into tokens to protect the query privacy. Otherwise, if the query includes some sensitive information, such as "$5 \leq priority \leq 7$", then the CS_2 could know the requester is querying some important users' metering data. Then, the requester or the query results could be traced or analyzed by the curious server CS_2.

4.2.3 Designing Goal

To enable effective range query over encrypted metering data under the aforementioned model, our design goal is to develop a privacy-preserving range query scheme over encrypted data for smart grid, and to achieve the security of the data and efficient range query as follows.

- The security requirements should be guaranteed in the proposed scheme. As stated above, if the smart grid does not consider the security, the residential users' privacy could be disclosed, and the real-time power metering reports could be stolen. Therefore, the proposed scheme should achieve the data confidentiality and privacy, as well as the query privacy.
- The performance efficiency should be achieved in the proposed scheme. As range query are operated over encrypted multidimensional data, compared with existing schemes, the proposed PaRQ scheme should improve the communication, computation and response time complexities.

4.3   The PaRQ Scheme

In this section, we show the details of the PaRQ. There are three major phases in our scheme: construction of the range query predicate phase, encrypted data deposit phase and range query phase. Firstly, we present the construction of the range query predicate phase.

4.3.1   Construction of the Range Query Predicate

Inspired by the equality predicate and comparison predicate, we can extend them to support range query predicate. Specifically, we can achieve the opposite semantics of the above comparison query, i.e., $x_i \leq j$, by constituting the vectors $\sigma(\mathbf{x})$ in a reverse manner as Eq. (4.1), since some part are presented in Chap. 1, thus, here is (4.1)

$$\sigma_{i,j} = \begin{cases} 1, \text{if } x_i \leq j, \\ 0, \text{otherwise.} \end{cases} \tag{4.1}$$

Thus, the HVE scheme can support range queries, such as $a \leq x_i \leq b$. Table 4.1 illustrates notations used in this chapter. The key generation phase is same as above in the quality query.

In the data encryption phase, the residential user should define two encryption vectors: $\sigma_{\geq}(\mathbf{x})$ and $\sigma_{\leq}(\mathbf{x})$ as Eqs. (1.2) and (4.1) when $x_i \geq j$ and $x_i \leq j$, respectively. The receiver can obtain the correct data if and only if both conditions $x_i \geq a$ and $x_i \leq b$ hold. If the encrypted data in HVE is Ω, the residential user $U_i \in \mathbf{U}$ should split Ω into two parts by the following steps: (1) randomly chooses a polynomial $f(x) = a'x + \Omega$, where a' is a random coefficient. (2) U_i chooses two random integers and computes two data shares Ω_L and Ω_R, i.e., Ω is divided into two parts: Ω_L and Ω_R. U_i encrypts Ω_L and Ω_R under vectors $\sigma_{\geq}(\mathbf{x})$ and $\sigma_{\leq}(\mathbf{x})$, respectively.

Table 4.1 The notations used in this paper

$\sigma(\mathbf{x})$	An encryption vector which related to the ciphertexts
$\sigma^*(\mathbf{w})$	A query vector which related to the query tokens
$s(\sigma^*(\mathbf{w}))$	A set of all indexes k where $\sigma^*(w_k) \neq *$
$\sigma_{\geq}(\mathbf{x})$	An encryption vector under the predicate $x_i \geq j$
$\sigma_{\leq}(\mathbf{x})$	An encryption vector under the predicate $x_i \leq j$
$\sigma_{\geq}^*(\mathbf{w})$	A query vector of a range query where $w_i \geq a$
$\sigma_{\leq}^*(\mathbf{w})$	A query vector of a range query where $w_i \leq b$
$s(\sigma_{\geq}^*(\mathbf{w}))$	A set of all indexes k where $\sigma_{\geq}^*(w_k) \neq *$
$s(\sigma_{\leq}^*(\mathbf{w}))$	A sets of all indexes k' where $\sigma_{\leq}^*(w_{k'}) \neq *$

In the token generation phase, the requester's range query are defined with two vectors: $\sigma_{\geq}^*(\mathbf{w})$ and $\sigma_{\leq}^*(\mathbf{w})$ when $w_i = a$ and $w_i = b$, respectively. Let $s(\sigma_{\geq}^*(w))$ be the set of all indexes k which satisfies $\sigma_{\geq}^*(w_k) \neq *$, and $s(\sigma_{\leq}^*(w))$ be the sets of all indexes k' which satisfy $\sigma_{\leq}^*(w_{k'}) \neq *$. Here, $k, k' \in (1, \ldots, nl)$. Finally, in the data query phase, the server checks two comparison predicates $P_{\sigma_{\geq}^*(\mathbf{w})}(\sigma_{\geq}(\mathbf{x}))$ and $P_{\sigma_{\leq}^*(\mathbf{w})}(\sigma_{\leq}(\mathbf{x}))$, which can be generated as Eq. (1.4). The server can obtain Ω_{L} if $\sigma_g(x_k)$ and $\sigma_{\geq}^*(w_k)$ are equal for all $k \in s(\sigma_{\geq}^*(w))$, i.e., $P_{\sigma_{\geq}^*(\mathbf{w})}(\sigma_{\geq}(\mathbf{x})) = 1$. Similarly, the server can obtain Ω_{R} if $\sigma_{\leq}(x_{k'})$ and $\sigma_{\leq}^*(w_{k'})$ are equal for all $k' \in s(\sigma_{\leq}^*(w))$, i.e., $P_{\sigma_{\leq}^*(\mathbf{w})}(\sigma_{\leq}(\mathbf{x})) = 1$. The range query predicate can be denoted as follows:

$$P_{(\sigma_{\geq}^*(\mathbf{w}), \sigma_{\leq}^*(\mathbf{w}))}(\sigma_{\geq}(\mathbf{x}), \sigma_{\leq}(\mathbf{x}))$$

$$= \begin{cases} 1, \text{if } P_{\sigma_{\geq}^*(\mathbf{w})}(\sigma_{\geq}(\mathbf{x})) = 1, and, P_{\sigma_{\leq}^*(\mathbf{w})}(\sigma_{\leq}(\mathbf{x})) = 1 \\ 0, \text{otherwise.} \end{cases} \quad (4.2)$$

Finally, if $P_{(\sigma_{\geq}^*(\mathbf{w}), \sigma_{\leq}^*(\mathbf{w}))}(\sigma_{\geq}(\mathbf{x}), \sigma_{\leq}(\mathbf{x})) = 1$ the server can recover Ω_{L} and Ω_{R}. Then, the data Ω can be computed.

The main procedures of range query on encrypted data in smart grid are illustrated in Fig. 4.3. The CC is not only the data forwarder but also the query translator. In the encrypted data deposit phase, before outsourcing his data, a residential user U_i encrypts his data m into a ciphertext C by randomly choosing a secret session key ks. At the same time, U_i hides ks and m's searchable attributes into another ciphertext CT by using the HVE range query predicate and the CC's pubic key PK. Note that, $\Omega = ks$ in our PaRQ. Then, U_i deposits both ciphertexts C and CT to the CC. The CC adds an index Ind to both C and CT. Then the CC transmits $\{Ind, \mathsf{C}\}$ to the CS_1 and $\{Ind, \mathsf{CT}\}$ to the CS_2.

As shown in Fig. 4.3, when a requester S posts a range query, the query should be translated into query tokens by using the CC's private key. Then, the requester deposits its tokens to the CS_2 to retrieve the session key ks and index Ind. The session keys whose encryption vectors are satisfied with the range query vectors and their indexes can be released to the requester. The requester queries the corresponding ciphertext C from CS_1 by using its received index Ind. Then, the requester can recover the original data by using the secret key ks to decrypt C.

4.3.2 The Encrypted Data Deposit Phase

4.3.2.1 Key Generation

For a single-authority smart grid system, a trusted authority (TA) can bootstrap the whole system. Specifically, in the key generation phase, given the security parameters κ, TA first generates $(q, g, \mathbf{G}_1, \mathbf{G}_2, e)$ by running $\mathscr{G}en(\kappa)$. TA randomly

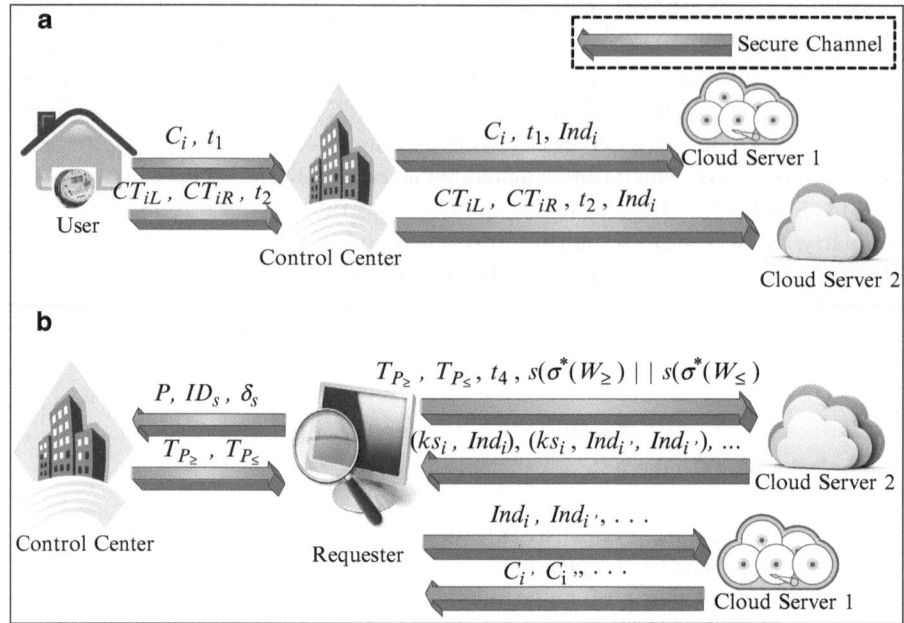

Fig. 4.4 Data flow in our PaRQ scheme. (**a**) Encrypted data deposit phase. (**b**) Range query phase

chooses a master key $r \in Z_q^*$, and computes the corresponding public key g^r. Thus, (g^r, r) is the public/private key pair of the TA. When S applies a query, TA assigns an ID-based key pair $(H_1(ID_S), H_1^r(ID_S))$, denoted as (pk_s, sk_s), to S. TA selects some random elements $g_1, g_2, (h_1, u_1, \psi_1), \ldots, (h_{nl}, u_{nl}, \psi_{nl}) \in \mathbf{G}_1$. The TA also picks random numbers $y_1, y_2, v_1, \ldots, v_{nl}, t_1, \ldots, t_{nl} \in \mathbf{Z}_p$. Then, TA computes $Y_1 = g^{y_1}, Y_2 = g^{y_2}, v_k = g^{v_k} \in \mathbf{G}_1$ for $k \in (1, \ldots, nl)$. In addition, TA computes $\Gamma = e(g_1, Y_1)e(g_2, Y_2) \in \mathbf{G}_2$. Later, TA distributes the HVE public/private key pair (PK, SK) to the CC as follows:

$$PK = (g, Y_1, Y_2, (h_1, u_1, \psi_1, V_1, T_1), \ldots, (h_{nl}, u_{nl}, \psi_l, V_{nl}, T_{nl}))$$

$$SK = (g_1, g_2, y_1, y_2, v_1, \ldots, v_{nl}, t_1, \ldots, t_{nl}).$$

We assume that the communication channels are secure in our system model. An ID-based signature scheme $Sig(\cdot)$ [8] can be used to authenticate the requester's identity. The details of secure channel establishment is without the scope of this paper. Figure 4.4 shows the dataflow of the PaRQ.

4.3.2.2 Data Encryption

We denote each data as m_i. When U_i wants to report m_i to the cloud server CS_1, U_i randomly generates a session key ks_i. Then U_i encrypts its data into a ciphertext CT_i, where $\mathsf{CT}_i = \mathsf{Enc}_{ks_i}(m_i)$. $\mathsf{Enc}(\cdot)$ is a symmetric encryption algorithm, e.g., AES [9].

For each uploading interval

$$U_i \rightarrow CC : \{\mathsf{C}_i, t_1\}$$

In this paper, "$A \rightarrow B : \{C\}$" means "A sends C to B". Then, the CC adds a unique index Ind_i to the data ciphertext and transmits all of them to the CS_1.

$$CC \rightarrow CS_1 : \{\mathsf{C}_i, t_1, Ind_i\}$$

4.3.2.3 HVE-Based Session Key Encryption

If each data has l searchable attributes, U_i chooses a vector $\mathbf{x}_i = (x_{i1}, \ldots, x_{il})$ $\in \sum^l$ to characterize its data m_i in different dimensions. To encrypt ks_i by using the CC's PK and the vector $\mathbf{x}_i$, U_i divides each ks_i into two parts: ks_{iL} and ks_{iR}. Then, ks_{iL} is encrypted by using the encryption vector $\sigma_\geq(\mathbf{x}_i)$; ks_{iR} is encrypted by using the encryption vector $\sigma_\leq(\mathbf{x}_i)$. Thus, the CS_2 can recover ks_i only when both encryption vectors are satisfied with the corresponding query vectors in the range query tokens. The HVE-based session key encryption details are as follows:

1. Firstly, U_i maps $\mathbf{x}_i$ to an encryption vector $\sigma_\geq(\mathbf{x}_i)$ as Eq. (1.2). Then, U_i selects two random numbers $r_{i1}, r_{i2} \in Z_p$ and computes tags for the ciphertext ks_{iL} by using the encryption vector $\sigma_\geq(\mathbf{x}_i)$ as:

$$\mathsf{C}_{L1}^i = Y_1^{r_{i1}}, \mathsf{C}_{L2}^i = Y_2^{r_{i1}},$$

$$\mathsf{C}_{L3,1}^i = (h_1 u_1^{\sigma_\geq(x_{i1})})^{r_{i1}} V_1^{r_{i2}},$$

$$\ldots,$$

$$\mathsf{C}_{L3,nl}^i = (h_{nl} u_{nl}^{\sigma_\geq(x_{inl})})^{r_{i1}} V_{nl}^{r_{i2}},$$

$$\mathsf{C}_{L4,1}^i = \psi_1^{r_{i1}} T_1^{r_{i2}},$$

$$\ldots,$$

$$\mathsf{C}_{L4,nl}^i = \psi_{nl}^{r_{i1}} T_{nl}^{r_{i2}},$$

$$\mathsf{C}_{L5}^i = g^{r_{i2}}, \mathsf{C}_{L6}^i = \Gamma^{r_{i1}} ks_{iL}.$$

$$(4.3)$$

Let $\mathsf{CT}_{iL} = (\mathsf{C}_{L1}^i, \mathsf{C}_{L2}^i, \mathsf{C}_{L3,1}^i, \ldots, \mathsf{C}_{L3,nl}^i, \mathsf{C}_{L4,1}^i, \ldots, \mathsf{C}_{L4,nl}^i, \mathsf{C}_{L5}^i, \mathsf{C}_{L6}^i)$.

2. Secondly, U_i maps $\mathbf{x}_i$ to an encryption vector $\sigma_{\leq}(\mathbf{x}_i)$ as Eq. (4.1). Then, U_i selects two random numbers $r'_{i1}, r'_{i2} \in Z_p$, and computes tags for the ciphertext $ks_{i\mathsf{R}}$ by using the encryption vector $\sigma_{\leq}(\mathbf{x}_i)$:

$$\mathsf{C}^i_{\mathsf{R}1} = Y_1^{r'_{i1}}, \mathsf{C}^i_{\mathsf{R}2} = Y_2^{r'_{i1}},$$

$$\mathsf{C}^i_{\mathsf{R}3,1} = (h_1 u_1^{\sigma_{\leq}(x_{i1})})^{r'_{i1}} V_1^{r'_{i2}},$$

$$\dots,$$

$$\mathsf{C}^i_{\mathsf{R}3,nl} = (h_{nl} u_{nl}^{\sigma_{\leq}(x_{inl})})^{r'_{i1}} V_{nl}^{r'_{i2}},$$

$$\mathsf{C}^i_{\mathsf{R}4,1} == \psi_1^{r'_{i1}} T_1^{r'_{i2}},$$

$$\dots,\quad (4.4)$$

$$\mathsf{C}^i_{\mathsf{R}4,nl} = \psi_{nl}^{r'_{i1}} T_{nl}^{r'_{i2}},$$

$$\mathsf{C}^i_{\mathsf{R}5} = g^{r'_{i2}}, \mathsf{C}^i_{\mathsf{R}6} = \Gamma^{r'_{i1}} ks_{i\mathsf{R}}.$$

Let $\mathsf{CT}_{i\mathsf{R}} = (\mathsf{C}^i_{\mathsf{R}1}, \mathsf{C}^i_{\mathsf{R}2}, \mathsf{C}^i_{\mathsf{R}3,1}, \dots, \mathsf{C}^i_{\mathsf{R}3,nl}, \mathsf{C}^i_{\mathsf{R}4,1}, \dots, \mathsf{C}^i_{\mathsf{R}4,nl}, \mathsf{C}^i_{\mathsf{R}5}, \mathsf{C}^i_{\mathsf{R}6}).$

4.3.2.4   Ciphertext Deposit

U_i deposits $\mathsf{CT}_{i\mathsf{L}}$ and $\mathsf{CT}_{i\mathsf{R}}$ to the CC as:

$$U_i \rightarrow CC : \{\mathsf{CT}_{i\mathsf{L}}, \mathsf{CT}_{i\mathsf{R}}, t_2\}.$$

The CC also adds the index Ind_i to the key ciphertext and transmits all of them to the CS_2.

$$CC \rightarrow CS_2 : \{\mathsf{CT}_{i\mathsf{L}}, \mathsf{CT}_{i\mathsf{R}}, t_2, Ind_i\}$$

4.3.3   Range Query Phase

4.3.3.1   Token Generation

When S wants to query the server to retrieve the expected data, S firstly generates a range query, such as $P = (a_1 \leq x_1 \leq b_1) \wedge (a_2 \leq x_2 \leq b_2) \dots \wedge (a_l \leq x_l \leq b_l)$. S then computes a signature $\delta_s = Sig_{sk_s}(P)$.

In each querying interval

$$S \rightarrow CC : \{P, ID_S, \delta_s\}.$$

The CC verifies S's signature by using S's public key. If S is an authorized requester, the CC might issue query tokens to S by following steps. The CC divides P into two parts: $P_\geq = (x_1 \geq a_1) \wedge (x_2 \geq a_2) \ldots \wedge (x_l \geq a_l)$ and $P_\leq = (x_1 \leq b_1) \wedge (x_1 \leq b_2) \ldots \wedge (x_1 \leq b_l)$. Let $w_\geq = (a_1, \ldots, a_l)$ and $w_\leq = (b_1, \ldots, b_l)$.

The CC generates two query vector $\sigma_\geq^*(\mathbf{w})$ and $\sigma_\leq^*(\mathbf{w})$ to represent $P_\geq$ and $P_\leq$ as Eq. (1.3), respectively. The wildcard $*$ in the vector $\sigma_\geq^*(\mathbf{w})$ means that S does not care about the attributes related to $*$. Let $s(\sigma^*(w_g))$ be the set of k which satisfies $\sigma_\geq^*(w_k) \neq *$. Let $s(\sigma_\leq^*(w))$ be the set of k' which satisfies $\sigma_\leq^*(w_{k'}) \neq *$. Then the CC computes a token $T_{P_\geq}$ by using the query vector $\sigma_\geq^*(\mathbf{w})$ as follows:

1. Select a random $\alpha, \beta \in Z_p$, and generate $\lambda_k, \psi_k, \gamma_k, \tau_k \in Z_p$ such that $\lambda_k y_1 + \psi_k y_2 = \alpha$, $\gamma_k y_1 + \tau_k y_2 = \beta$ for all $k \in s(\sigma_\geq^*(w))$.
2. Compute a token $T_{P_\geq}$ as

$$K_{L1}^s = g_1 \prod_{k \in s(\sigma_\geq^*(w))} (h_k u_k^{\sigma_\geq^*(w_k)})^{\lambda_k} \psi_k^{\gamma_k},$$

$$K_{L2}^s = g_2 \prod_{k \in s(\sigma_\geq^*(w))} (h_k u_k^{\sigma_\geq^*(w_k)})^{\varphi_k} \psi_k^{\tau_k},$$

$$K_{L3}^s = g^\alpha, \tag{4.5}$$

$$K_{L4}^s = g^\beta,$$

$$K_{L5}^s = g^{-\sum_{k \in s(\sigma_\geq^*(w))} (v_k \alpha + t_k \beta)}.$$

Similarly, the CC generates a token $T_{P_\leq}$ by using the query vector $\sigma_\leq^*(\mathbf{w})$ as follows:

1. Select a random $\alpha', \beta' \in Z_p$, and generate $\lambda_{k'}, \psi_{k'}, \gamma_{k'}, \tau_{k'} \in Z_p$ such that $\lambda_{k'} y_1 + \psi_{k'} y_2 = \alpha'$, $\gamma_{k'} y_1 + \tau_{k'} y_2 = \beta'$ for all $k' \in s(\sigma_\leq^*(w))$.
2. Compute the token $T_{P_\leq}$ as

$$K_{R1}^s = g_1 \prod_{k' \in s(\sigma^*(w_{ls}))} (h_{k'} u_{k'}^{\sigma^*(w_{lsk'})})^{\lambda_{k'}} \psi_{k'}^{\gamma_{k'}},$$

$$K_{R2}^s = g_2 \prod_{k' \in s(\sigma^*(w_{ls}))} (h_{k'} u_{k'}^{\sigma^*(w_{lsk'})})^{\varphi_{k'}} \psi_{k'}^{\tau_{k'}},$$

$$K_{R3}^s = g^{\alpha'}, \tag{4.6}$$

$$K_{R4}^s = g^{\beta'},$$

$$K_{R5}^s = g^{-\sum_{k' \in s(\sigma^*(w_{ls}))} (v_{k'} \alpha' + t_{k'} \beta')}.$$

Let $T_{P_\geq} = (K_{L1}^s, K_{L2}^s, K_{L3}^s, K_{L4}^s, K_{L5}^s)$ and $T_{P_\leq} = (K_{R1}^s, K_{R2}^s, K_{R3}^s, K_{R4}^s, K_{R5}^s)$. Then, the CC keeps a record $(ID_s, T_{P_\geq}, T_{P_\leq})$ in its database and distributes $T_{P_\geq}$ and $T_{P_\leq}$ to the requester S as its authorized tokens:

$$CC \rightarrow S : \{T_{P_\geq}, T_{P_\leq}\}$$

4.3.3.2   Key and Index Query

After receiving the query tokens from the CC, the requester deposits them as well
as the non-wildcard indexes sets to the cloud server CS_2 as:

$$S \rightarrow CS_2 : \{T_{P_{\geq}}, T_{P_{\leq}}, t_4, s(\sigma^*_{\geq}(w)), s(\sigma^*_{\leq}(w))\}$$

Then, the CS_2 searches its database to find whether there is a key ciphertext which
matches the requester's query conditions. For each key ciphertext, if its encryption
vectors are satisfied with the query vectors in the query tokens, the CS_2 can obtain:

$$ks_{i_L} = \frac{D_1.D_2.e(K^s_{L5}, C^i_{L5}).C^i_{L6}}{e(K^s_{L1}, C^i_{L1}).e(K^s_{L2}, C^i_{L2})}, \tag{4.7}$$

$$ks_{i_R} = \frac{D'_1.D'_2.e(K^s_{R5}, C^i_{R5}).C^i_{R6}}{e(K^s_{R1}, C^i_{R1}).e(K^s_{R2}, C^i_{R2})}, \tag{4.8}$$

where $D_1 = e(K^s_{L3}, \prod_{k \in s(\sigma^*_{\geq}(w))} C^i_{L3,k})$ and $D_2 = e(K^s_{L4}, \prod_{k \in s(\sigma^*_{\geq}(w))} C^i_{L4,k})$. $D'_1 =$
$e(K^s_{R3}, \prod_{k' \in s(\sigma^*_{\leq}(w))} C^i_{R3,k'})$ and $D'_2 = e(K^s_{R4}, \prod_{k' \in s(\sigma^*_{\leq}(w))} C^i_{R4,k'})$. Thus, the CS_2
can obtain ks_i by using ks_{i_L} and ks_{i_R}. Then, the CS_2 sends this recovered key ks_i
with indexes to the requester.

$$CS_2 \rightarrow S : \{(ks_i, Ind_i), (ks_{i_,}, Ind_{i'}), \ldots\}$$

Note that, there might be more than one session keys which are satisfied with
the range query tokens. Then, the CS_2 distributes all of session keys back to the
requester S.

4.3.3.3   Data Query

Upon receiving the keys and indexes from the CS_2, the requester queries the CS_1
by using the received indexes to obtain the corresponding data ciphertext.

$$S \rightarrow CS_1 : \{Ind_i, Ind_{i'}, \ldots\}$$

Then, the CS_1 searches its database to find whether there are ciphertexts matching
the requester's indexes. If so, the CS_1 sends matched ciphertexts to the requester S:

$$CS_1 \rightarrow S : \{C_i, C_{i'}, \ldots\}$$

4.3.3.4   Data Decryption

After receiving the real session keys from the CS_2 and ciphertexts $\{\mathsf{C}_i, \mathsf{C}_{i'}, \ldots\}$ from the CS_1, the requester S can obtain the real data by using the session keys to decrypt the ciphertexts, otherwise, C_i can be discarded.

$$S : m_i = \mathsf{Dec}_{ks_i}(\mathsf{C}_i), \ldots$$

$\mathsf{Dec}(\cdot)$ is the symmetric decryption algorithm corresponding to the opposite operation of $\mathsf{Enc}(\cdot)$.

4.3.4   Enhancement with Collusion Resilience

In our system model, we assume that the adversary cannot control both CSs. Actually, in order to prevent the cloud server CS_1 and CS_2 in collusion to disclose the data m_i, an identity based encryption scheme [10] can be used in the data encryption phase to encrypt m_i. For instance, the CC has one pair of identity based public/private key (pk, sk). Firstly, m_i is encrypted by the CC's public key pk. Again it is encrypted by using the session key ks_i and be outsourced to the CS_1. Thus, even if the CS_2 recovers the session key ks_i with the requester's query tokens, the CS_2 can not decrypt the ciphertexts on the CS_1 to obtain the data m_i. The reason is that the CS_2 cannot obtain the CC's ID-based private key sk. When an authorized requester S asks for query tokens from the CC, the CC replies the query tokens as well as its identity based private key sk. Accordingly, the requester can recover data m_i by using both session key ks_i from CS_1 and sk from the CC. Note that, the requesters are high-level users, such as energy company's financial auditors, who are authorized by the CC. Therefore, they can obtain the CC's private key to decrypt their queried data. As a result, the PaRQ can remain secure even the two cloud servers are in collusion.

Furthermore, to provide forward security and prevent requesters from decrypting future encrypted data by using CC's old private key, the CC can compute different pairs of time and identity based public/private key pairs at different time intervals. Therefore, the authorized requesters can only obtain CC's private keys corresponding to their entitled intervals to decrypt their required data.

4.4   Security Analysis

In this section, we analyze the security properties of the proposed PaRQ according to the security requirements discussed in Sect. 4.2.

- *The individual residential users' data confidentiality can be achieved.* In the PaRQ, the residential user's data m_i is encrypted by its session key ks_i. For the eavesdroppers and the CS_2, they cannot obtain anything from the ciphertext C_i because they lack of the secret key ks_i. Although the CS_2 can recover ks_i from the requester's query tokens, the CS_2 cannot extract the ciphertexts from the CS_1 if they are not in collusion. Our enhancement introduced in section V.D can be resilient to the collusion attack even if the CS_1 is in collusion with the CS_2. Accordingly, the individual residential user's data confidentiality is achieved in the proposed PaRQ scheme.
- *The individual residential users' data privacy can be preserved.* In our PaRQ, ks_i and the searchable attributes are hid in two ciphertexts $\{CT_{iL}, CT_{iR}\}$. Other requesters cannot obtain ks_i from the CS_2 if they cannot obtain the authorized query tokens $\{T_{P_\geq}, T_{P_\leq}\}$ from the CC or their query vectors in the query tokens are not satisfied with the encryption vectors on the ciphertexts. Accordingly, they cannot pry into the session keys and decrypt the encrypted metering data. The range query correctness can be demonstrated as follows. In Eq. (4.7), the numerator equals:

$$D_1 \cdot D_2 \cdot e(K_{L5}^s, C_{L5}^i) \cdot C_{R6}^i$$

$$= e\left(g^\alpha, \prod_{k \in s(\sigma_\geq^*(w))} (h_k u_k^{\sigma_\geq(x_{ik})})^{r_{i1}} g^{v_k r_{i2}} \right)$$

$$\cdot e\left(g^\beta, \prod_{k \in s(\sigma_\geq^*(w))} \psi_1^{r_{i1}} g^{t_1 r_{i2}} \right)$$

$$\cdot e\left(g^{-\sum_{k \in s(\sigma_\geq^*(w))} (v_k \alpha + t_k \beta)}, g^{r_{i2}} \right) \cdot \Gamma^{r_{i1}} ks_{iL}$$

$$= e\left(g^\alpha, \prod_{k \in s(\sigma_\geq^*(w))} (h_k u_k^{\sigma_\geq(x_{ik})})^{r_{i1}} \right)$$

$$\cdot e\left(g^\beta, \prod_{k \in s(\sigma_\geq^*(w))} \psi_k^{r_{i1}} \right) \cdot e\left(g^{r_{i2}}, \prod_{k \in s(\sigma_\geq^*(w))} g^{v_k \alpha + t_k \beta} \right)$$

$$\cdot e\left(g^{-\sum_{k \in s(\sigma_\geq^*(w))} (v_k \alpha + t_k \beta)}, g^{r_{i2}} \right) \cdot \Gamma^{r_{i1}} ks_{iL}$$

$$= e\left(g^\alpha, \prod_{k \in s(\sigma_\geq^*(w))} (h_k u_k^{\sigma_\geq(x_{ik})})^{r_{i1}} \right)$$

$$\cdot e\left(g^\beta, \prod_{k \in s(\sigma_\geq^*(w))} \psi_k^{r_{i1}} \right) \cdot \Gamma^{r_{i1}} ks_{iL}. \tag{4.9}$$

while the denominator equals:

$$e(K_{L1}^s, C_{L1}^i).e(K_{L2}^s, C_{L2}^i)$$

$$= e\left(g_1 \prod_{k \in s(\sigma_{\geq}^*(w))} (h_k u_k^{\sigma_{\geq}^*(w_k)})^{\lambda_k} \psi_k^{\gamma_k}, g^{y_1 r_{i1}}\right)$$

$$\cdot e\left(g_2 \prod_{k \in s(\sigma_{\geq}^*(w))} (h_k u_k^{\sigma_{\geq}^*(w_k)})^{\varphi_k} \psi_k^{\tau_k}, g^{y_2 r_{i1}}\right)$$

$$= \Gamma^{r_{i1}} . \prod_{k \in s(\sigma_{\geq}^*(w))} [e((h_k u_k^{\sigma_{\geq}^*(w_k)})^{\lambda_k}, g^{y_1 r_{i1}})$$

$$\cdot e((h_k u_k^{\sigma_{\geq}^*(w_k)})^{\varphi_k}, g^{y_2 r_{i1}})]$$

$$\cdot \prod_{k \in s(\sigma_{\geq}^*(w))} [e(\psi_k^{\gamma_k}, g^{y_1 r_{i1}}).e(\psi_k^{\tau_k}, g^{y_2 r_{i1}})]$$

$$= \Gamma^{r_{i1}} . \prod_{k \in s(\sigma_{\geq}^*(w))} e((h_k u_k^{\sigma_{\geq}^*(w_k)})^{r_{i1}}, g^{\lambda_k y_1 + \varphi_k y_2})$$

$$\cdot \prod_{k \in s(\sigma_{\geq}^*(w))} e(\psi_k^{r_{i1}}, g^{\gamma_k y_1 + \tau_k y_2})$$

$$= \Gamma^{r_{i1}} . e\left(\prod_{k \in s(\sigma_{\geq}^*(w))} (h_k u_k^{\sigma_{\geq}^*(w_k)})^{r_{i1}}, g^{\alpha}\right)$$

$$\cdot e\left(\prod_{k \in s(\sigma_{\geq}^*(w))} \psi_k^{r_{i1}}, g^{\beta}\right) \tag{4.10}$$

Let Θ be the set of indexes $i \in s(\sigma_{\geq}^*(w))$ where $\sigma_{\geq}^*(w_k) \neq \sigma_{\geq}(x_{ik})$. The Eq. (4.7) outputs follows

$$\frac{D_1.D_2.e(K_{s50}, C_{i50}).C_{i60}}{e(K_{s10}, C_{i10}).e(K_{s20}, C_{i20})}$$

$$= e(g^{\alpha}, \prod_{k \in \Theta} (u_k^{\sigma_{\geq}(x_{ik}) - \sigma_{\geq}^*(w_k)})^{r_{i1}}).ks_{iL} \tag{4.11}$$

$$= e(g, g)^{\alpha r_{i1} \Sigma_{k \in \Theta}(\log_g(u_k))(\sigma_g(x_{ik}) - \sigma_{\geq}^*(w_k))}$$

If $\sigma_{\geq}^*(w_k)$ equals $\sigma_{\geq}(x_{ik})$ for all $k \in s(\sigma_{\geq}^*(w))$, ks_{iL} can be recovered; otherwise, the unauthorized requesters cannot obtain ks_{iL} according to Eq. (4.7).

Similarly, the unauthorized requesters cannot obtain ks_{i_R} from Eq. (4.8). Therefore, only the authorized requester can obtain the query results and the users' data privacy is preserved.

- *The requester's query privacy can be preserved.* In our PaRQ, a requester's query is divided into two parts: $P_\geq$ and $P_\leq$, by the CC. Both of them are translated into tokens $T_{P_\geq}$ and $T_{P_\leq}$ in the form of $\prod_{k \in s(\sigma_\geq^*(w))}(h_k u_k^{\sigma_\geq^*(w_k)})^{\lambda_k} \psi_k^{\gamma_k}$ and $\prod_{k' \in s(\sigma_\leq^*(w))}(h_{k'} u_{k'}^{\sigma_\leq^*(w_{k'})})^{\varphi_{k'}} \psi_{k'}^{\tau_{k'}}$, respectively. For the eavesdroppers, they can learn nothing about the query P only with the tokens $T_{P_\geq}$ and $T_{P_\leq}$ because they have no idea about the index sets and encryption parameters $\lambda_k, \psi_k, \gamma_k, \tau_k$. Since the CS_2 still does not know the encryption parameters $\lambda_k, \psi_k, \gamma_k, \tau_k$, it also cannot obtain the real value of P even with the tokens and the index sets $s(\sigma_\geq^*(w))$ and $s(\sigma_\leq^*(w))$. Therefore, the requester's query privacy is preserved in the proposed PaRQ scheme.

Table 4.2 Comparison of security properties

Properties	PaRQ [11]	MRQED [12]	Buket [13]	OPE [14]
Confidentiality	Yes	Yes	Yes	Yes
Data privacy	Yes	Yes	Partial	Partial
Query privacy	Yes	No	No	No

From the above security analysis and comparison in Table 4.2, our PaRQ can achieve all of the data confidentiality and privacy and query privacy, compared with order-preserving encryption (OPE) [14] based technique and bucketization (Buket) based technique [13].

4.5 Performance Evaluation

In this section, we evaluate the performance of the proposed PaRQ scheme in terms of the communication overhead, computation complexity and response time of the system.

4.5.1 Communication Overhead

We numerically analyze the communication overhead of our PaRQ, compared with the MRQED [12], in terms of the public key size, ciphertexts size and token size. Since the functionality of the decryption key computation phase in determining the query results in MRQED is similar to that of query tokens in the PaRQ, therefore, we take their size in comparison. "Tok/Dek" in Tables 4.3 and 4.4 represents Token or Decryption key. Since most pairing-based cryptosystems need to work in a subgroup of the elliptic curve $E(F_q)$, by representing elliptic curve points using point compression, the lengths of the elements in G_1 and G_2 are roughly 161-bit (using point compression) and 1,024-bit, respectively. In the following, l is the number of data dimensions and N is domain of attribute values.

In the key generation phase, the public key includes $(5Nl + 3)$ G_1 elements. If we choose AES ciphertext with 256-bit, the data ciphertext C_i is only of 256 bits. In addition, session key's ciphertext includes two parts: CT_{iL} and CT_{iR}, each of which includes $(2Nl + 3)$ G_1 elements and a G_2 element, thus the size of each session key's ciphertext is $(4Nl + 6) \times 161 + 2,048$ bits. Since each query token includes 5 G_1 elements, the size of the two query tokens $\{T_{P_\geq}, T_{P_\leq}\}$ is $161 \times 10 = 1,610$ bits.

Table 4.3 Comparison of communication complexity (bit)

Communication	PaRQ	MRQED [12]
Public key	$(5Nl + 3) \times 161$	$8Nl \times 161 + 1,024$
Ciphertext	$(4Nl + 6) \times 161 + 2,304$	$(4Nl + 1) \times 161 + 1,024$
Tok/Dek	1,610	$5Nl \times 161$

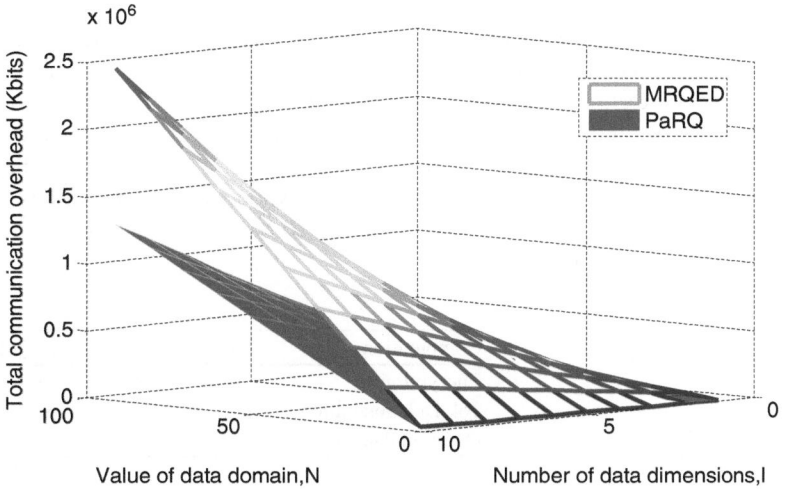

Fig. 4.5 Comparison of communication overhead between PaRQ and MRQED schemes

In comparison, the public key in the MRQED [12] includes $8Nl$ G_1 elements and a G_2 element. The decryption keys include $5Nl$ G_1 elements. In addition, there are $(4Nl + 1)$ G_1 elements and a G_2 element in the ciphertexts. Table 4.3 shows that compared with the MRQED, our PaRQ consumes less communication overhead. Especially, our PaRQ significantly reduces the tokens transmission overhead, which is a constant, i.e., 1,600 bits; in the MRQED the transmission overhead of the decryption keys may increase with both l and N. The total communication overhead comparison is depicted in Fig. 4.5. It further indicates that our PaRQ costs less communication overhead than the MRQED.

4.5.2 Computation Overhead

In our PaRQ, the computation tasks include pairing operations and exponentiation operations. For simplicity of description, the pairing operation and exponentiation operation are denoted as C_p and C_e, respectively. Since the AES encryption/decryption and multiplication are much faster than the pairing operations, we do not analyze the AES encryption/decryption and multiplication in this subsection.

Fig. 4.6 Tandem model [15] of a range query process

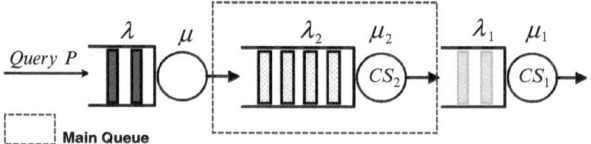

For the PaRQ, the symmetric encryption of C_i is very fast. Meanwhile, the corresponding session key is encrypted into key ciphertexts $\{CT_{i0}, CT_{i1}\}$ by using its encryption vectors. The computation overhead of $\{CT_{i0}, CT_{i1}\}$ is $(10Nl + 8)C_e$ because each part requires $(5Nl + 4)$ exponentiation operations. In the token generation phase, the computation cost of $\{T_{P_\geq}, T_{P_\leq}\}$ is $(12l + 2)C_e$, because each query token in $\{T_{P_\geq}, T_{P_\leq}\}$ needs $6l + 3$ exponentiation operations. After receiving the tokens, the CS_2 needs to compute 10 pairings to recover the session key $\{ks_{iL}, ks_{iR}\}$, i.e., $10C_p$.

Table 4.4 Comparison of computation complexity

Computation	PaRQ	MRQED [12]
Encryption	$(10Nl + 8)C_e$	$(8Nl + 3)C_e$
Tok/Dek generation	$(12l + 6)C_e$	$8NlC_e$
Query in database	$10C_p$	$5l \cdot log NC_p$

On the other hand, the MRQED [12] needs $(8Nl + 3)$ exponentiation operations to encrypt a message, another $8Nl$ exponentiation operations to derive the decryption keys and $5l \cdot log N$ pairing operations to search the correct results. From Table 4.4, we can see that the encryption overhead in both PaRQ and MRQED increase with l and N. The computation overhead of token generation in the PaRQ only increases with l, whereas, the overhead of decryption key generation in MRQED increases with both l and N. When a query is executed in a database, the overhead in our PaRQ is a constant ($10C_p$); the overhead in the MRQED still increases with both l and N. Hence, our PaRQ is much more efficient than the MRQED. Further comparison of their range query response time is analyzed in following subsection.

4.5.3 Response Time

To provide good services to requesters, the response time of a range query is an important metric. For example, it would be useful for the requesters to know how long they exactly need to wait for a range query result so that they can efficiently schedule their tasks. Actually, response time varies according to many factors, such as communication latency etc. We analyze the response time of our PaRQ with or without considering the network communication latency Δ. The other factors are not being included in this calculation of the response time.

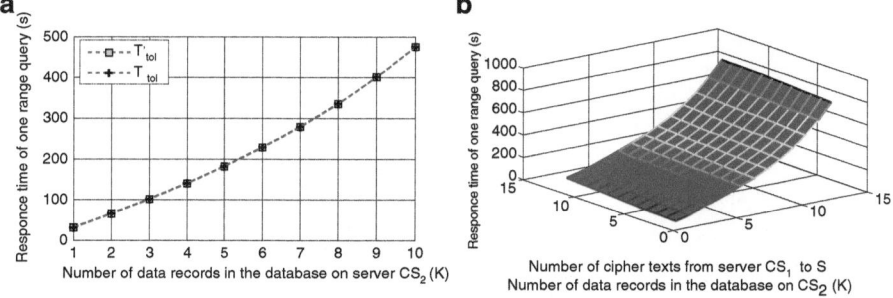

Fig. 4.7 Response time in PaRQ scheme. (**a**) $\Delta = 0$. (**b**) $\Delta = 0.022 + 4.26t$

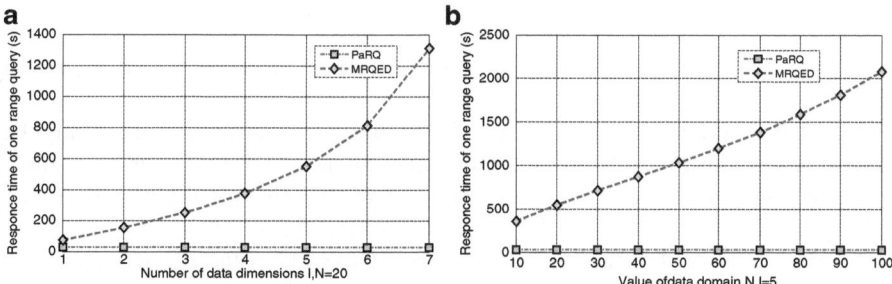

Fig. 4.8 Comparison of the number of matched residential users, $1/\mu_2 = 3.1$ s. (**a**) $\lambda_2 = \lambda_2' = 1/600, 1/\mu_2' = 0.6699l$. (**b**) $\lambda_2 = \lambda_2' = 1/600, 1/\mu_2' = 0.775 \times log_2(N)$

In the PaRQ, a range query is processed by the CC, CS_1 and CS_2. We model our range query process as a tandem model of network queues [15], as shown in Fig. 4.6.

We assume that the range query arrives the system according to a poisson process with rate λ, and uses the CC for token generation in an exponentially distributed time interval with mean $1/\mu$ (as an $M/M/1$ queue). Upon exiting the CC, the requester continue accesses the CS_2 with rate λ_2 for a time which is deterministic $1/\mu_2$ (as an $M/D/1$ queue). Finally, the requester accesses server CS_1 with rate λ_1 for a time which is exponentially distributed with mean $1/\mu_1$ (as an $M/M/1$ queue). Let

$$\rho = \frac{\lambda}{\mu}; \rho_1 = \frac{\lambda_1}{\mu_1}; \rho_2 = \frac{\lambda_2}{\mu_2}.$$

If all the network states are in the set of $n = \{n_0, n_1, n_2\}$, according to Jackson's Theorem [16], the steady-state probability distribution of the system is given as:

$$P(n_1, n_2, n_3) = \rho^{n_0}(1 - \rho)\rho_1^{n_1}(1 - \rho_1)\rho_2^{n_2}(1 - \rho_2).$$

Let T, T_1 and T_2 be the average queuing delay of the CC, CS_1 and CS_2, respectively. They can be calculated as:

$$T = \frac{1}{\mu - \lambda} = \frac{1}{\mu(1 - \rho)};$$

$$T_1 = \frac{1}{\mu_1 - \lambda_1} = \frac{1}{\mu_1(1 - \rho_1)};$$

$$T_2 = \frac{1}{\mu_2}\frac{2 - \rho_2}{2 - 2\rho_2}.$$

Then, the total delay of the range query in the PaRQ is:

$$T_{tol} = \frac{1}{\mu(1 - \rho)} + \frac{1}{\mu_1(1 - \rho_1)} + \frac{1}{\mu_2}\frac{2 - \rho_2}{2 - 2\rho_2}.$$

In this section, the response time of a range query is the total queuing delay on all the servers.

Detailed experiments are conducted on a Pentium IV 3-GHz system to study the execution time [17]. For G_1 over the Freeman-Scott-Teske (FST) curve, a single exponentiation operation in G_1 with 161 bits costs 1.1 ms, and the corresponding pairing operation costs 3.1 ms. Without loss of generality, let $N = 20, l = 5$. According to Table 4.4, the processing time of the CC is the tokens generation time, i.e., $1/\mu = (12l + 6) \times 1.1 = 72.6\,\text{ms} \approx 0.073\,\text{s}$. If range query length is exponentially distributed with mean 2 Kbits and arrives according to a poisson process with rate 1 query/10 min, i.e., $\lambda = 1/600$, and the queuing delay $T = \frac{1}{\mu - \lambda} = 0.073\,\text{s}$ and average queue length $L = \frac{\lambda}{\mu - \lambda} = 0.0012$.

Next, the processing time of the CS_2 depends on the query tokens verification and the searching time in CS_2's database. The computation time of a query token verification over one record is $10 \times 3.1 = 31$ ms. If the number of records in CS_2's database is $M = 100$, the CS_2's processing delay is 3.1 s, i.e. $1/\mu_2 = 3.1$ s. At last, querying indexed ciphertexts on the CS_1 is typically processed very fast, the processing time is only several milliseconds (e.g. $1/\mu_1 = 10$ ms). Since $\mu_2 = max(\lambda_1)$, considering the extreme case $\lambda_1 = \mu_2$, then, $T_1 = 0.01$ s and $L_1 \approx 0$. Therefore, T, L, T_1, L_1 are very small, and few rang query tasks can be buffered in the queue CS_1 and CC. As a result, their queuing delay can be neglected.

From the above analysis, the processing time on the CC is much faster than the query arriving interval. Thus, $\lambda = \lambda_2 = 1/600$. Moreover, compared with the CS_2 queue, the service rate of the CS_1 and CC is much faster, i.e., $\mu_1 \gg \mu \gg \mu_2$. Therefore, a range query's average response time is mainly determined by the processing time of the CS_2. Consequently, the whole range query response time is distributed approximately as in the CS_2 queue, that is, as in an $M/D/1$ queue with poisson rate λ_2 and service rate $1/\mu_2$. Hence, the response time of a range query can be represented by T'_{tol}.

$$T'_{tol} = \frac{1}{\mu_2} \frac{2 - \rho_2}{2 - 2\rho_2}.$$

If the total communication latency Δ among all network links is not negligible, the formula T'_{tol} should be adjusted to

$$T'_{tol} = \frac{1}{\mu_2} \frac{2 - \rho_2}{2 - 2\rho_2} + \Delta.$$

In fact, the smart grid usually uses 3G (3rd Generation) or 4G (4th Generation) cellular network topology for cells data transmission. Data transmission rate is 60–240 Kbps, and distance converge depends upon the availability of cellular service [18]. Hence, the communications rate among the requesters, the CC and the CSs in our PaRQ scheme is assumed to be 240 Kbps. Here, $\lambda_2 = 1/600$ and $1/\mu_2 = 3.1$ s.

1. If the communication latency Δ is negligible, i.e., $\Delta \approx 0$, we can see from Fig. 4.7 that the response time of a range query is increased with the number of the database records. Comparing the total response time of a range query with or without considering the queuing delay of the CC and CS_1 by using T_{tol} and T'_{tol}, respectively, Fig. 4.7a also shows that they are almost the same, which means that the queuing delay of the CS_1 and CC really can be neglected in response time calculation.

2. If the communication latency Δ is not negligible, we should consider the communication overhead during the range query process. From the above analysis, if S sends a 2 K-bit range query to the CC, the CC replies with two 1,610-bit tokens. Then, S forwards these two tokens to the CS_2, and the CS_2 replies the satisfactory keys and indexes. If t is the average number of matched results and the size of the keys and indexes are 80 bits each, thus, their communication overhead is

160t bits. Finally, when S accesses the CS_1, the CS_1 replies the correct indexed ciphertexts. Usually, the size of the ciphertext is large. Let it be 1 Mbits/packet. The communication overhead of the ciphertexts is 1t Mbits. Hence, the system communication latency $\Delta \approx 0.022 + 4.26t$.

$$T'_{tol} = \frac{1}{\mu_2}\frac{2-\rho_2}{2-2\rho_2} + 0.022 + 4.26t.$$

Figure 4.7b illustrates that both the number of data records in the CS_2's database and the number of replied ciphertexts from the CS_1 can augment the system response time.

3. Compared with the MRQED, Table 4.2 shows that the MRQED needs $5l \cdot log\,NC_p$ to verify a query on a ciphertext. In a database with the number of data records $M = 100$, the processing time in the database is $5l \cdot log\,N \times 3.1$ s. Similarly, in the MRQED, the encryption and decryption key generation time is much shorter than the processing time of data query in the database. Therefore, the range query response time of the MRQED is mainly determined by the query in database. Thus, the range query service in the MRQED can also be modeled by using an $M/D/1$ queue with poisson rate $\lambda'_2 = 1/600$ and service rate $1/\mu'_2 = 5l \cdot log\,N \times 3.1$. Figure 4.8 illustrates the response time comparison between the PaRQ and the MRQED without communication delay, the range query arrival rates are $\lambda_2 = \lambda'_2 = 1/600$. From Table 4.3 and Fig. 4.8, we can observe that in the MRQED, if the number of data records in the database is a constant, the response time of a range query increases with the domain N and the number of dimensions l, while the service time in our PaRQ is $1/\mu_2 = 3.1$ s. Thus, no matter how many dimensions of the data and how large the domain is, a single range query process time in the PaRQ remains $\frac{1}{\mu_2}\frac{2-\rho_2}{2-2\rho_2} = 3.1$ s, which is much less than that of the MRQED.

4.6   Related Works

4.6.1   Security and Privacy in Smart Grid

Security and privacy are critical to the development of wireless networks, especially for the real-time data audit strategy in smart grid. The smart grid interpretability panel-cyber security working group [3] presents some guidelines for smart grid cyber security, including security strategy, architecture, and high-level requirements. Li et al. [7] reviews the cyber security and privacy issues in smart grid and discusses some security and privacy solutions for smart grid. Lu et al. [2] use a super-increasing sequence to structure multidimensional data and encrypt the structured data by the holomorphic paillier cryptosystem technique. Li et al. [19] propose an authentication scheme based on Merkle tree for smart grid. Acs and Castelluccia

[20] exploit the privacy-preserving aggregation technique of time-series data in smart meters. They employ a differential privacy model in which users add noise to their electricity metering and the aggregator can successfully obtain the sum of the metering with a very large probability. In summary, few works focus on the query, especially range query over encrypted data in smart grid, which is really significant for user's metering data audit.

4.6.2 Range Query

Recently, the problem of querying encrypted data has been deeply investigated in both cryptography and database communities. One of the widely studied approaches is public key encryption with keyword search (PEKS) [21]. PEKS can protect users' data privacy and certain query privacy. However, most of PEKS schemes, such as the Searchable Encryption Scheme for Auction (SESA) [22], only can be applied for equality checks. Range query over the encrypted data with numeric attributes is more difficult, and most of the existing literatures cannot achieve data and query privacy simultaneously.

Roughly speaking, there are four categories of solutions that have been developed for range queries: order-preserving encryption (OPE), bucketization (Bucket), HVE and special data structure traversal. OPE-based technique [14] is to ensure that the order of plaintext data is preserved in the ciphertext domain. This allows direct translation of range predicate from the original domain to the domain of the ciphertext. However, the coupling distribution of plaintext and ciphertext domains might be exploited by attackers to guess the scope of the corresponding plaintext for a ciphertext [23]. Bucket-based technique [13] uses distributional properties of the datasets to partition and index data for efficient querying while trying to keep the information disclosure to a minimum. Queries are evaluated in an approximate manner where the returned set of records may contain some false positives.

In an HVE-based approach [24], two vectors over attributes are associated with a ciphertext and a token, respectively. Under the predicate translator, the ciphertext matches the token if and only if the two vectors are component-wise equal. Several HVE schemes [25–27] have been proposed in literatures. All of them use bilinear groups equipped with bilinear maps, and each constructs a proper method to hide attributes in an encrypted vector. However, it is expensive to compute exponentiation and pairing in a composite-order group. Jong [25] proposes a new HVE scheme that not only works in prime-order groups, but also requires a shorter token size and fewer pairing computations. However, Jong's scheme cannot be directly applied in the smart grid applications where data are high in dimension, variety or both.

Some specialized data structured for range query evaluation are trying to preserve notions of semantic security of the encrypted data, such as B+ tree etc. Recently, Shi et al. [12] propose a searchable encryption scheme that supports multidimensional range queries over encrypted data (MRQED). The MRQED utilizes an interval tree structure to form a hierarchical representation of intervals for each dimension and stores multiple ciphertexts corresponding to a single data value on the server, i.e., each one corresponds to a range. If it is applied to a single-dimensional data with values belonging to a domain of size N. The ciphertext representation is $O(logN)$ times the actual data. If the MRQED is applied to a piece of data with l dimensions, each query requires l times complexity to execute.

4.7 Summary

In this chapter, we have proposed a privacy-preserving range query scheme, named PaRQ, for smart grid. An HVE based range query predicate is constructed to realize the range query on encrypted metering data. The PaRQ allows users to store their data on cloud servers in encrypted form, and range queries can be executed by using cloud server's computational capabilities. A requester with authorized query tokens can obtain the correct session keys to retrieve the metering data within specific query ranges. Security analysis demonstrates that the PaRQ can achieve data confidentiality and privacy and preserve query privacy. Performance evaluation shows that the PaRQ can significantly reduce computation and communication overhead, as well as response time. For our future work, we intend to enhance our PaRQ to support ranked range query with security and privacy preservation.

References

1. R. Zeng, Y. Jiang, C. Lin, and X. Shen, "Dependability analysis of control center networks in smart grid using stochastic petri nets," *IEEE Transactions on Parallel and Distributed Systems*, vol. 23, no. 9, pp. 1721–1730, 2012.
2. R. Lu, X. Liang, X. Li, X. Lin, and X. Shen, "Eppa: An efficient and privacy-preserving aggregation scheme for secure smart grid communications," *IEEE Transactions on Parallel and Distributed Systems*, vol. 23, no. 9, pp. 1621–1631, 2012.
3. The Smart Grid Interoperability Panel-Cyber Security Working Group, "Nistir 7628 guidelines for smart grid cyber security: Smart grid cyber security strategy, architecture, and high-level requirements." http://csrc.nist.gov/publications/nistir/ir7628/nistir-7628_vol1.pdf, August 2010.
4. X. Liang, X. Li, R. Lu, X. Lin, and X. Shen, "Udp: Usage-based dynamic pricing with privacy preservation for smart grid.," *IEEE Transactions on Smart Grid*, vol. 4, no. 1, pp. 141–150, 2013.
5. R. Yu, Y. Zhang, S. Gjessing, C. Yuen, S. Xie, and M. Guizani, "Cognitive radio based hierarchical communications infrastructure for smart grid," *IEEE Network*, vol. 25, no. 5, pp. 6–14, 2011.
6. C. Wang, Q. Wang, K. Ren, and W. Lou, "Privacy-preserving public auditing for data storage security in cloud computing," in *Proc. INFOCOM*, pp. 1–9, 2010.
7. X. Li, X. Liang, R. Lu, X. Shen, X. Lin, and H. Zhu, "Securing smart grid: cyber attacks, countermeasures, and challenges," *IEEE Communications Magazine*, vol. 50, no. 8, pp. 38–45, 2012.

8. B. Libert and J.-J. Quisquater, "The exact security of an identity based signature and its applications.," *IACR Cryptology ePrint Archive*, vol. 2004, p. 102, 2004.
9. J. Daemen, V. Rijmen, and A. Proposal, "Rijndael," in *Proc. AESCC*, 1998.
10. D. Boneh and M. Franklin, "Identity-based encryption from the weil pairing," in *Proc. CRYPTO*, pp. 213–229, Springer, 2001.
11. M. Wen, R. Lu, K. Zhang, J. Lei, X. Liang, and X. Shen, "Parq: A privacy-preserving range query scheme over encrypted metering data for smart grid," *IEEE Transactions on Emerging Topics in Computing*, vol. 1, no. 1, pp. 178–191, 2013.
12. E. Shi, J. Bethencourt, T. Chan, D. Song, and A. Perrig, "Multi-dimensional range query over encrypted data," in *Proc. SP*, pp. 350–364, 2007.
13. B. Hore, S. Mehrotra, M. Canim, and M. Kantarcioglu, "Secure multidimensional range queries over outsourced data," *The International Journal on Very Large Data Bases*, vol. 21, no. 3, pp. 333–358, 2012.
14. A. Boldyreva, N. Chenette, and A. O'Neill, "Order-preserving encryption revisited: Improved security analysis and alternative solutions," in *Proc. CRYPTO*, pp. 578–595, Springer, 2011.
15. D. Bertsekas, R. Gallager, and P. Humblet, *Data networks*, vol. 2. Prentice-Hall International, 1992.
16. J. Walrand, "A probabilistic look at networks of quasi-reversible queues," *IEEE Transactions on Information Theory*, vol. 29, no. 6, pp. 825–831, 1983.
17. M. Scott, "Efficient implementation of cryptographic pairings," in *[Online]. http://www.pairing-conference.org/2007/invited/Scottslide.pdf*, 2007.
18. P. Parikh, M. Kanabar, and T. Sidhu, "Opportunities and challenges of wireless communication technologies for smart grid applications," in *Proc. PESGM*, pp. 1–7, 2010.
19. H. Li, X. Liang, R. Lu, X. Lin, and X. Shen, "Edr: an efficient demand response scheme for achieving forward secrecy in smart grid," in *Proc. GLOBECOM*, pp. 929–934, IEEE, 2012.
20. G. Acs and C. Castelluccia, "I have a dream!(differentially private smart metering)," in *Proc. IH*, pp. 118–132, Springer, 2011.
21. D. Boneh, G. Di Crescenzo, R. Ostrovsky, and G. Persiano, "Public key encryption with keyword search," in *Proc. Eurocrypt*, pp. 506–522, Springer, 2004.
22. M. Wen, R. Lu, J. Lei, H. Li, X. Liang, and X. S. Shen, "Sesa: an efficient searchable encryption scheme for auction in emerging smart grid marketing," *Security and Communication Networks*, vol. 7, no. 1, p. 234–244, 2013.
23. R. Agrawal, J. Kiernan, R. Srikant, and Y. Xu, "Order preserving encryption for numeric data," in *Proc. SIGMOD*, pp. 563–574, ACM, 2004.
24. D. Boneh and B. Waters, "Conjunctive, subset, and range queries on encrypted data," in *Proc. TCC*, pp. 535–554, 2007.
25. J. Park, "Efficient hidden vector encryption for conjunctive queries on encrypted data," *IEEE Transactions on Knowledge and Data Engineering*, vol. 23, no. 10, pp. 1483–1497, 2011.
26. J. Katz, A. Sahai, and B. Waters, "Predicate encryption supporting disjunctions, polynomial equations, and inner products," in *Proc. EUROCRYPT*, pp. 146–162, Springer, 2008.
27. V. Iovino and G. Persiano, "Hidden-vector encryption with groups of prime order," in *Proc. Pairing*, pp. 75–88, Springer, 2008.

Chapter 5
Conclusions and Future Works

5.1 Conclusions

In this brief, we have investigated querying techniques in the smart grid. We conclude the brief with the following remarks.

1. In the smart grid, two-way transmission creates large scale of data which come from sensors, individual archives, social networks, internet of things, enterprise and internet in all scales and formats. The most challenging issue is how to effectively manage such a large amount of data and identify efficient ways to analyze these data and unlock information.
2. One important aspect in the smart grid is the data security and privacy issue. Many works have been carried out, focusing on attack detection, data mining and analysis, etc. However, data privacy issue is seldom mentioned. Due to its extraordinary scale, data privacy can be leaked at the encryption stage, transmission stage and data querying stage in the smart grid. Though, some literatures explore data encryption and data aggregation to achieve the goal, there still exist many challenges, such as efficient encryption and decryption algorithms, encrypted data indexing techniques.
3. The requirements for market analysis and decision support, for predicting future expected power demand and prices, for monitoring in case of unexpected behavior, make it necessary to query over encrypted data, such as equality query, conjunctive query and range query.
4. Querying techniques over encrypted data are different from querying over relational database. Many existing database indexing technologies cannot be used in this scenario because that the encrypted data should not be decrypted when indexing or querying over them.
5. In this brief, we focus on how to query over encrypted data in the smart grid for auction in energy market and financial auditing etc. The multidimensional characteristics of the data is also considered. Both the data privacy and data confidentiality are protected in our proposed query techniques.

5.2 Future Research Directions

In this brief, we have investigated the problem of preserving data privacy and query privacy in the smart grid. Our solutions are mainly based on the keyword search over encrypted data in the smart grid. Compared to the various query functions supported by the relational database, our query solutions can be further extended to support ranked query, fine-grained query with logical operations and disjunctive query etc.

For example, in the smart grid auction market, energy buyers may want to filter out the energy with reasonable price by using the range of price keyword. In addition, ranked query results among multidimensional keywords can provide buyers more flexible choices. Thus, our proposed SESA scheme, which is introduced in the Chap. 2, can be extended to achieve this goal. Further, the range query with ranked search can be combined with an efficient aggregation algorithm to reduce the communication overhead.

At the big data era, users in the smart grid may want to achieve fine-grained query over encrypted data, by giving multi-keyword search with logical operations, i.e., AND, OR and NO. Therefore, how to provide more comprehensive query functionalities is a challenging issue. If the relevance of keywords is considered, precise keyword query could be achieved. Furthermore, if the weight of keywords is considered, the users could get better query results.

Finally, compared to conjunctive query, there also exists disjunctive query. How to achieve disjunctive query over encrypted data is never considered by any existing literatures. Therefore, it would be valuable in exploring disjunctive query over encrypted data by analysing three kinds of output streams (the true-, the false-, and the unknown-streams). They are generated for a selection or join-node, accordingly, a CNF (conjunctive normal form)- or DNF (disjunctive normal form)- based query predicate can be constructed to resolve the problem. In addition, CNF- or DNF-based encryption predicate also can be explored to support the disjunctive query over encrypted data.